Immersion Into Noise

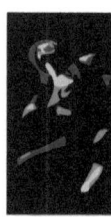
Critical Climate Change

SERIES EDITORS: TOM COHEN AND CLAIRE COLEBROOK

The era of climate change involves the mutation of systems beyond 20th century anthropomorphic models and has stood, until recently, outside representation or address. Understood in a broad and critical sense, climate change concerns material agencies that impact on biomass and energy, erased borders and microbial invention, geological and nanographic time, and extinction events. The possibility of extinction has always been a latent figure in textual production and archives; but the current sense of depletion, decay, mutation and exhaustion calls for new modes of address, new styles of publishing and authoring, and new formats and speeds of distribution. As the pressures and re-alignments of this re-arrangement occur, so must the critical languages and conceptual templates, political premises and definitions of 'life.' There is a particular need to publish in timely fashion experimental monographs that redefine the boundaries of disciplinary fields, rhetorical invasions, the interface of conceptual and scientific languages, and geomorphic and geopolitical interventions. Critical Climate Change is oriented, in this general manner, toward the epistemo-political mutations that correspond to the temporalities of terrestrial mutation.

Immersion Into Noise
Joseph Nechvatal

OPEN HUMANITIES PRESS

London, 2022

Second edition published by Open Humanities Press 2022
First edition published by Open Humanities Press 2011
Freely available online at http://openhumanitiespress.org/immersion-into-noise

Copyright © 2011 Joseph Nechvatal, New Preface Copyright © 2022 Joseph Nechvatal
This is an open access book, licensed under the Creative Commons By Attribution Non-Commercial Share Alike license. Under this license, authors allow anyone to download, reuse, reprint, modify, distribute, and/or copy this book so long as the authors and source are cited, the use is not commercial, and resulting derivative works are licensed under the same or similar license. Read more about the license at creativecommons.org/licenses/by-nc-sa/3.0

Cover image: Joseph Nechvatal, maelstrOm autOmata 2011, computer-robotic assisted acrylic on velvet, 50 x 50 cm, Galerie Richard, Paris

ISBN Print: 978-1-78542-124-2
ISBN PDF: 978-1-78542-123-5

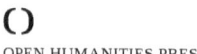

OPEN HUMANITIES PRESS

OPEN HUMANITIES PRESS is an international, scholar-led open access publishing collective whose mission is to make leading works of contemporary critical thought freely available worldwide. More at http://openhumanitiespress.org

Contents

List of Figures   7

Preface   9

The Art of Noisy Noologies   13

Into Noise: Tabula Rasa vs. Horror Vacui   35

Noise Vision   59

Signal-to-Noise Eye   104

Modern Nervous Noise Eyes   133

Viral Attack within Connectivist Noise Schematics   199

Noise Against Oblivion:
An Omnijective Philosophy of Noise Culture   209

Notes   230

Bibliography   258

Additional Licenses   269

List of Figures

Figure 1: Uplifting, 1983, 11x14", graphite on paper, Joseph Nechvatal

Figure 2: Palace of Power, 1984, 11x14" graphite on paper, Joseph Nechvatal

Figure 3: XS the Opera: Shakespeare Theatre Boston 1986

Figure 4: Enhanced detail image from the Abside of the Grotte de Lascaux, Dordogne (France)

Figure 5: Gods of Politics, 1984, 14x11" graphite on paper, Joseph Nechvatal

Figure 6: Side ossuary in the Cimitero dei Cappuccini located beneath the chapel Santa Maria della Concezione (Rome)

Figure 7: Detail of a section of the Parisian Catacombs (Paris)

Figure 8: Rosario Chapel in Santo Domingo Church (Puebla, Mexico)

Figure 9: Rococo interior of the Ottobeuren Abbey (Bavaria)

Figure 10: Interior view of Egid Quirin Asam's Asamkirche (Munich)

Figure 11: Fidelis Schabet's décadent Venus Grotto, 1877 (Linderhof)

Figure 12: Interior stairway of Victor Horta's Hôtel Tassel (Brussels)

Figure 13: Exterior view of Antoni Gaudí's Casa Batlló (Barcelona)

Figure 14: Exterior view of the Palais Idéal by Facteur Cheval (Hauterives, Drôme, France)

Figure 15: Hiroshima after the dropping of the atomic bomb

Preface

"No longer considered only a factor of disturbance, detrimental to information like static noise in the channel of communication, the evolving concept of noise also becomes constitutive of new forms of knowledge and of new ways of understanding organization."
—Cécile Malaspina, *An Epistemology of Noise*

"I wanted to swallow myself by opening my mouth very wide and turning it over my head so that it would take in my whole body, and then the Universe, until all that would remain of me would be a ball of eaten thing which little by little would be annihilated: that is how I see the end of the world."
—Jean Genet, *Our Lady of the Flowers*

From classical myth to contemporary culture, the palimpsest of artful noise has fascinated and troubled the imagination. A decade has expired since the 2011 first edition publication of *Immersion Into Noise*. Since then, posh cultural observers have (clutching their pearls) increasingly asserted that smart phone addicted people have reduced even further their tolerance for (and interest in) long-form difficult cultural expressions. The hyper-pop ethos—quick, simple, clear, swipe and move on— is assumed to be an inevitably slippery downward slide into the lowest common denominator of Populism. The intolerably pedantic assumption is that technological pop life no longer provides the means for the appreciation of poetic art. Noise in this swift, smooth, abysmally vapid decline, represents the boldest of *bête noires*.

Behind this gassy glossolalia lay all the complexities of silence. That is why the art-of-noise, since 2011, has been taking on increasingly powerful importance by remaining anti-doctrinaire. It is normal that marginal, counter-cultural expressions gain import as their frequency declines. In 2022, the art-of-noise is the phantom remaining myth in the absence of mythical memory. With noise art, we do vital art. Pop culture is what is done to us—to be intuitively enjoyed, but eluded.

I think it still cuts the mustard, but since the first edition of *Immersion Into Noise* I have had a few revelations that may give the book an added contextual reverberation. Perhaps it should have been more scornful towards celebrity and identity (both forms of clarity), though it was not short on sensually provocative flair and perverse audacity. Writing it, I wanted to be a nimble noisy theorist, not a noise theorist *manqué*. For noise is at some level always a political concept. So I understand the book today less as oracular, and more as a miasma of clattering sly burlesque that focused a bit heavily on the poetic formative aesthetics of post-punk pestilence. I certainly romanticized decay as a basis for an avuncular post-apocalyptic passion, fueled by the honey blood of dead magic. But I am proud that the book called for a more dynamic, tolerant, understanding, and multi-layered self.

I realize now that there is a thread of sybaritic paganism running through the book at odds with most printed reflections on the merits of noise as a form of cultural resistance. Indeed, I have a fondness for cultural noise that also has an aspect of sensuous luxury (or pleasure) about it. I know now that an early influence on this was watching The Three Stooges' slapstick film *Pop Goes the Easel* (1935) on TV at home in Chicago when I was around 11 years old. Two salient points: my mother had forbidden me watching any Three Stooges (thus a tasty transgression) and the art student clay slinging scene reminded me that when I was an infant, I 'painted' the white wall next to my crib with my feces.

But mostly, the wrinkled libertinage ideas in *Immersion Into Noise* need no re-calibration for this second edition. I continue to practice them as an art-of-noise artist, though I would stress harder the palimpsest psychic divinational function of noise dissimulation, and less the endless lacuna.

With the viral pandemic, *cri de coeur* cultural communications of the noisy *viractual* kind (my 1999 portmanteau for the interface between the virtual and the actual) have only increased. I myself have produced two long, semi-purloined, noise music pieces for the Pentiments experimental music and sound art label. Besides releasing my retrospective *Selected Sound Works* (1981-2021) tracks on cassette and on the internet at Bandcamp, Pentiments has produced a double vinyl LP of my recent

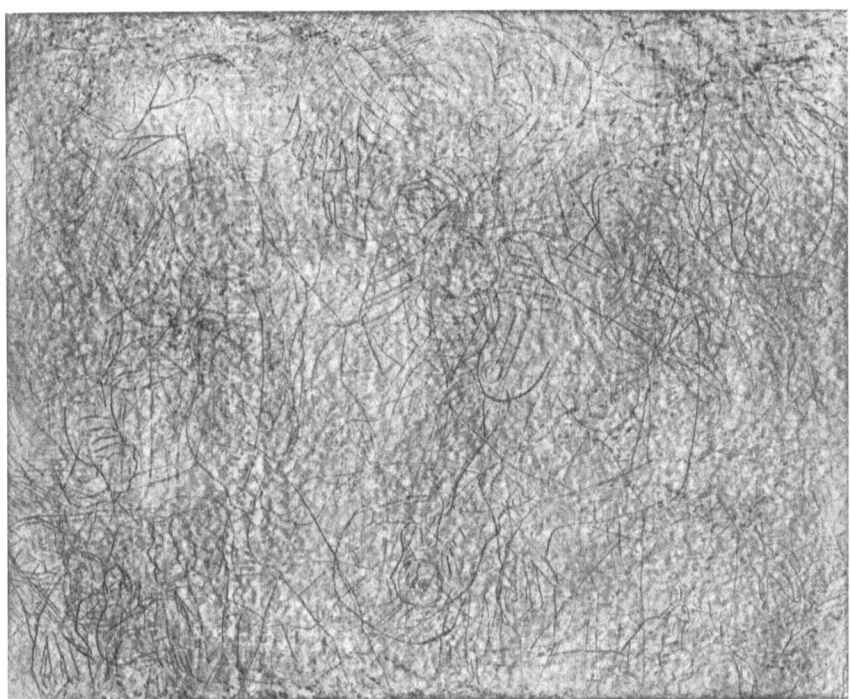

Figure 1 – Uplifting, 1983, 11 x 14", graphite on paper, Joseph Nechvatal. In the collection of the Frederick R. Weisman Art Museum, University of Minnesota

noise music called *The Viral Tempest*. One record contains *OrlandO et la tempête viral symphOny redux suite* (2020) and the other *pour finir avec le jugement de dieu viral symphOny plague* (2021). These long sonic works make use of a collapse between signal and ground in the same way I have continued to explore in my visual art compressing figure and ground—for example in my 2020 series of a-life virus-modeled paintings called *Orlando et la tempête* (Orlando and the Tempest). That art show was an invitation—extraordinarily poignant during the pandemic—to imagine being alive as pansexual other within an ambiguous viral storm that is whipping up a queerer, trans, multi-racial tolerant world of playful pleasure. It promoted that which is characterized in French as *caprice*.

That caprice compression almost defines the art-of-noise aesthetic for me. It also conveys something of the forlorn charm of the overlapping, shimmering chaos of the information society, where the figure is absorbed into excess without fixity; mercurial drifting within fields of nasty noise. But what else has changed in those ten years, is that, at first,

I hadn't quite realized the *haute volupté* diabolic glee of the art of noise. Noise is what blows through us. Its *outré* ~ off beat ~ tight tonal contrasts allow us to find our people by acknowledging the ghosts of our elders. That art of noise lineage—that tossing of pearls before swine—is what we have as culture that does not merely help us pass the time, but which enlivens it if we surrender to its intoxicating difficulty. We really have to see music, philosophy, art, and politics as sorts of separate languorous noise drones in constant interplay with one another. What we have to recognize in 2022—something I did not realize in 2011—is that the interplay between the different drones isn't a matter of one monitoring or reflecting another, but of exploring the uncertainties and anxieties about noise that takes place between them. It is that which points us towards the magical occult in our age of the passive picturesque.

Artistic deviance is always defined in terms of criteria of legitimacy or illegitimacy. In 2011 I tried to return you to yourself in the grandeur of a majestic and gorgeous ocean of noise. In 2022 I ask you to address *Immersion Into Noise*, rather than only being addressed by it. Because *Immersion Into Noise* is positioned in an aesthetic where corporate forms of display are confronted by a noisy semi-private vernacular. As in 2011, reading this book is an antagonistic exercise in mentally pulverizing pop culture in order to release into life the latent becoming of art: that becoming which makes art not only a sign-object, but also a lapidary intersubjective form of vanguard communication; if albeit draped in capricious cheap swag. But we came from raven stars, magnetic fields, and libertine noise fields of atoms and molecules—and that mystical throbbing is where we irretrievably are headed.

INTRODUCTION

The Art of Noisy Noologies

Where there is simple information processing, there is simple experience, and where there is complex information processing, there is complex experience.
— David Chalmers, "Facing up to the Problem of Consciousness"

[...] the value of an old work of art should be assessed on the basis of the amount of radical theory that can be drawn from it, on the basis of the nucleus of creative spontaneity which the new creators will be able to release from it [...]
— Raoul Vaneigem, The Revolution of Everyday Life

What is real is the becoming itself, the block of becoming, not the supposedly fixed terms through which that which becomes passes.
— Gilles Deleuze and Félix Guattari, A Thousand Plateaus

You never know what is enough unless you know what is more than enough.
— William Blake, The Marriage of Heaven and Hell

Applied to culture, the concept of immersion into noise is an unfamiliar one. What do I mean when I say *immersion*? It is by no means uncomplicated. It is, like art, gradient. Even the word *noise*[1] on its own is fraught with philosophical implications and deserves to be closely scrutinized. And what is this diaphanous, dark designation called *noise music* and—even more problematically—*art noise*?

It would be preposterous to pass judgement on the *aesthetics of noise* without asking such questions at the start. And so to begin, I will endeavor to answer them in terms of what I will propose as an *art of noise*.[2] This art of noise will be embodied in the literary/scholastic technique used here, one that embraces a noisy heterogeneity that at times may simultaneously be fun, frustrating and funny. Thus what I offer the reader is a text that will engage them in a dynamic play of noisy forces and fluctuating perspectives that exemplifies the propositions put forth here: propositions concerning the recognition of—and immersion into—*cultural noise*.

First, we must consider that noise takes place in a general media culture of massive electronic deluge, where the mercurial reproduction of free-floating (ineffable) signifiers of language, sound and images has blurred into a problematized complex/compound/prodigality sometimes referred to as *information overload*. In one respect, all sounds and images are already a kind of noise: data without meaning.[3] But I want to argue that an *art noise* takes a slight step outside overload/intoxication by means of its fuzzy identity as *art* (something vaguely abstract that is linked to pleasure and critique). Of course, any simplistic explication of the function of art (a concept that has no single function, but several) within Western society alone today would be inept. However, by examining the various definitions offered over the centuries, we can see that the *idea of art* developed primarily out of notions of anthropomorphic aesthetic agency displayed first through manual dexterity and then through intellectual stratagems concerning collective or intimate demonstration. As this embraces many types of production that are not conventionally deemed to be art, perhaps a better term for art would be *culture*.[4]

If so, perhaps an art of noise can also be postulated as a realm of antisocial cultural purpose directed toward the revolutionary transformation of an irrational social reality that insists on calling itself rational. I would like to think so and will argue this with the support of Gilles Deleuze's (1925-1995) notion of the *vacuole*. This concept of noncommunication comes from Deleuze's *Postscript on Control Societies*.[5] Deleuze's notion of control is connected to information-communication technology—a concept he pulled out of the work of William S. Burroughs (1914-1997). A vacuole is like a sac in a cell's membrane, completely bound up inside the cell but also separate from it. Vacuoles play a significant role in autophagy, maintaining an *imbalance* between biogenesis (production) and *degradation* (or turnover) of many substances and cell structures. They also aid in the destruction of invading bacteria or of misfolded proteins that have begun to build up within the cell. The vacuole is a major part of the plant and animal cell.

If we agree to combine this thought of *noise art as a vacuole of noncommunication* with an insistence on signal-to-noise[6] psychological circuit breaking, we gain a more complicated image of noise—as vacuoles that re-route and break-up the pathways of control. Let us therefore entertain a noncommunicating art of noise as an aesthetic act that nevertheless communicates intricately.

Consequently, I will focus here on the *beatific aspects of noise* (as I see it, the *negation of negation*) connected to an abstract non-communication as located in art that uses noise to re-route and break up our mental habits.[7] Thus the focus will be on signal-to-noise art relations, those relations that signal anti-social interruption, resistance, damage and frustration as sources of psychic pleasure. This concentration directs us towards an understanding of art noise as an art that distorts and disturbs crisp signals of cultural communications.

Our focus will conclude with a broader theory of pleasurable resistance as applied to *noise culture* in viral infected networked systems.

So to begin, I hypothesize that an art (or culture) of noise[8] produced in our milieu of image superabundance and information proliferation can problematize culture and hence enliven us to the privacy of the human condition *in lieu* of the fabulously constructed social spectacle[9] that engulfs and (supposedly)[10] controls us.

As a consequence, this book's aim is to open a dissonant space for a beatific noise theory that constitutes an alternative, although not necessarily a competitor, to the quiet manner in which most art and music theory is generally practised. As such, its strategic goal is focused less on delivering to the reader a sealed cultural product of recognition, and more on calling the reader into an immersive state of procedure that is based on the attributes of continuous spanning (*distentio*). This emphasis on the continuous spanning of listening—which itself is indicative of the immersive modus operandi of sound—lends a focus to thought that delivers a sense of continuity over time (*extentio animi*), as opposed to readily available—and thus fixed—intellectual strategies. This is particularly so in that the starting point of this intellectual investigation is the immersive position from within: *intus*. A position that necessitates a broad span of hearing, sight and thought—as well as a tight focus on disturbance.

Towards an Immersive Noise Consciousness

The highest art will be the one that presents in its contents of consciousness the thousand-fold problems of the time; to which one may note that this art allows itself to be tossed by the explosions of the last week, that it pieces together its parts again and again while being shoved by the day before.
— Tristan Tzara, Franz Jung, George Grosz, Marcel Janco, Richard Hulsenbeck, Gerhard Preiss, Raoul Hausmann, *Dada Manifesto*

...our previous history is not the petrified block of single visual space since, looked at obliquely, it can always be seen to contain its moment of unease.
— Jacqueline Rose, *Sexuality in the Field of Vision*

In everyday use, the word noise means unwanted sound or noise pollution. I look at it (and listen to it) differently: from an *immersive* perspective. In music, dissonance is the quality of sounds that seem *unstable*, with an aural "need" to "resolve" to a "stable" consonance. Despite the fact that words like "unpleasant" and "grating" are often used to describe the sound of harsh dissonance, in fact all music with a harmonic or tonal basis—even music that is perceived as generally harmonious—incorporates some degree of dissonance.

For music, I am using the term *immersion* in a strong sense: sound surrounds us, and in a weak sense: as a spontaneous substitution involved in suspending disbelief and outside stimulus for an interval of time—as when one's attention gets wrapped up in something compelling. For visual art, the term will be applied to suspending disbelief when using one's own interpretative imagination—plus a more physically based, and more scopic, application. For *consciousness*, the term is here almost mashed up with *immersion*.

As with art, reductive explanations of consciousness have proved impossible.[11] And as Marcel Duchamp (1887-1968) said, the fact that ready-mades are regarded with the same reverence as objects of art probably means he failed to solve the problem of trying to do away entirely with art. Thus, *Immersion Into Noise* will take on a very wide aesthetic interpretation of noise (in art) and push it to the limit: defining immersive noise art as a *saturating border experience*.[12] So the raucous understandings proposed here are going to fashion a synchronous theory of art, particularly informed by encounters with—and concepts of—the inside (and outside) of sensual noise. By attacking the important abstract aspects of aesthetic noise, *Immersion Into Noise* will propose a supplementary understanding of contemporary culture. But it will also touch on aspects of ancient Western culture as detected in the histories of art and architecture and so develop a general theory of *immersive noise consciousness*: one that is a *disturbing, sensorially reverberating, compound unified field*.

In electronics, noise can refer to the electronic signal corresponding to acoustic noise (in an audio system) or the electronic signal corresponding to the (visual) noise commonly seen as "snow" on a degraded television or video image. In signal processing (or computing) it can be considered data without meaning; that is, data that is not being used to transmit a signal, but is simply produced as an unwanted by-product of other activities. Noise can block, distort, or change the meaning of a message in both human and electronic communication. What the art of noise does is to take the meaninglessness of noise and convert it into the meaningful.

White noise is a random signal (or process) with a flat power spectral density. In other words, the signal contains equal power within a fixed bandwidth at any center frequency. White noise is considered analogous to white light, which contains all frequencies.

The French philosopher Michel Serres has interrogated the idea of noise in two of his books, *Genèse* and *The Parasite*, where he established that inherent to the concept of noise is the incident of *interference*. For him, noise is a chaotic parasite that is an excluded middle (or third)—without which the entire logical structure of western thought is unthinkable.[13]

In noise art, modes of representation (categories) tend to be interfered with and thus bend towards collapse. I intend to show how the cavernous conversion in aesthetic perception engendered by noise (as it wraps around us) can also be stretched to identify certain shifts in ontology that are relevant to our understanding of *being* by attending to sound-wave vibration frequency. To do so, an automatismic artistic-philosophical consideration of noise must assume the two-fold task of establishing an axiomatic aesthetic epistemology based on theoretical texts (of artists whenever possible), while testing them against my own artistic experiences and by placing myself within the operations of noise, thus raising questions of the reciprocity between theory and practice. I have approached this reciprocity through the creation of a 99.28 minute laptop noise opus entitled *viral symphOny* in four movements with *pOstmOrtem* (2008).[14]

Unsurprisingly, the popularity of glitch electronica[15] and its clicks-and-cuts aesthetic of error (and to a lesser extent *musique concrète*) directly relates to my interest in noise as a form of negative dialectics, as it mines what was once the erroneous use of musical technology in the production of sound. In glitch, the effects of malfunction, such as bugs, crashes, system errors, hardware noise,[16] skipping and audio distortion, can be captured on computers and this material provides the fundamentals of glitch music. In the glitch sense, deciphering noise in art will be tied to the potent erroneous in a general way. But noise art is not pop, and the broad spectrum of people do not appreciate it. In each case, from the mainstream point of view, something is wrong with the art. Subsequently, I will examine what noise signifies to those who love it.

To do so, I will be looking at the cultural and aesthetic benefits of noise[17] from my point of view both as a practising artist and as an art theoretician. Hence, in addition to preparing the reader for the previously indicated stylistic bounces back and forth between the first and the third person voice in the text, I shall establish my fundamental contention that all art is fundamentally conceptual and imaginative because art only exists conceptually and its goal is to change our consciousness. That is what Joseph Kosuth teaches us. This is not an uncomplicated matter however, for as the philosopher and specialist in consciousness studies, Dr. David Chalmers, says in his seminal essay "Facing up to the Problem of Consciousness", "there is nothing that we know more intimately than conscious experience, but there is nothing harder to define".[18]

According to Gilles Deleuze, consciousness is "the passage, or rather the awareness of the passage, from less potent totalities to more potent ones, and vice versa".[19] This hypothesis receives support from Thomas Metzinger who writes in *Conscious Experience* that "holism is a higher-order property of consciousness" and that "this global unity of consciousness seems to be the most general phenomenological characteristic of conscious experience".[20]

Noise is often loud, elaborate, interlaced and filigreed—but almost always gradient and highly phenomenological.[21] Noise is often a potent and transcendent negating intensity—but it is never unassimilable. The prime example of this will be the short history of noise music to follow.[22] What was once noise (unacceptable) has now become noise music (acceptable and even desirable).[23] What was once negating and exterior now fuels the inner artistic imagination. But for noise to be first noise, it must *destabilize* us. It must initially *jar*. It must *challenge*. It must initiate a glitch of psychic crumbling.

Through the relationship between noise and noise music[24] we can see how both notions of *in* and *out* (of the psychic edge/frame) are contained in an expanded idea of noise[25] that becomes potentially unhindered.[26] This understanding offers the critic a complex amplitude that deepens both the inside and the outside because they both extend as part of a potentially vast scope. In noise, potentially opposing directions can lose their positions and meet in a crushed connectivity. This recalls Gilles Deleuze's wonderful statement that "the interior is only a selected exterior, and the exterior, a projected interior".[27] If we amend these divisions within a *conceptual noise homogeneity* I believe we are much closer to the truth of the matter as concerns the experiential levels of aesthetic immersion into noise.

Noise art theory, then, involves the exaltation of the void and the melting of unstable frontiers as it expands definitions both inwardly and outwardly to envelope from both sides a felt understanding of the unfettered immensity and myrrh of our universe (where noise of one sort or another is everywhere).[28]

Given this thinking about noise's special conundrum (particularly when depending upon the supposed filter of analogic thought, the step-by-step comparison of partial similarities between things) *art as noise*—or *noise in art*—is well suited for reflecting on noise's overwhelming sensations and qualities of excess: an incoherent and multivalent excess that defeats attempts at reducing reality[29] to the indexical level of representation. I am theorizing here, then, a potentially shifting total excess where many once discrete elements are conceived as occupying the same space in a preliminary step towards producing an innovative unity (or ontological identity) as *mash chaosphere*.[30]

Undoubtedly, at times in the past totalizing analogies have been fatuously and unequivocally self-sententious in their urge towards perfectification, embellished (as they must seem to be) with a sort of self-significance and often fallacious, sweeping universalism. However, this awkward question of totalizing in terms of immersive noise art and immersive noise consciousness is at the hub of this investigation[31] and so it immediately beckons the questions: what models we make for ourselves, which do we prefer aesthetically, and why?

In terms of creativity and self-programming, the resultant new sense of synthesized unity noise provides is valuable in that it allows the creator to move from what exists and is known to the limits of knowledge and experience, and therefore to move into the realm of the unknown—a move from the familiar to the unconceived.

To address this gradient subject of audio noise—and then take it outside the organ of the ear—I have accumulated aesthetic examples of noise tendencies as experienced by myself and as found in the histories of art and philosophy. These examples subsequently contribute towards the articulation of what I have come to call *noise culture*—or better—*noise consciousness*,[32] as even the proclamation, *culture*, presents a set of highly ambiguous notions in that the word *culture* has immediately conflicting insinuations, and it is invariably best to observe scrupulously the context of its use. For some it means High Art, but for others (myself included) the word has more anthropological applications where culture represents less hypothetical measures of excellence than a widespread *way of hearing, seeing and being*.

What I will show here is that while formulating such an art of noise consciousness, a good deal of the basis for the questioning of the western ontological tradition has been found in the western tradition itself when we hear with new ears and look with new eyes and ask indeterminate questions.

Noisy Aesthetics

Whatever is fitted in any sort to excite the ideas of pain and danger, that is to say, whatever is in any sort terrible, or is conversant about terrible objects, or operates in a manner analogous to terror, is a source of the sublime; that is, it is productive of the strongest emotion which the mind is capable of feeling.

— Edmund Burke, *A Philosophical Enquiry into the Origin of Our Ideas of the Sublime and Beautiful*

The term *aesthetics* is traditionally used to distinguish the appreciative from the expedient. The notion originated with a 1739 text by the German philosopher Alexander Gottlieb Baumgarten (1714-1762) who introduced the term *aesthetic* in his text *Meditationes Philosophicae de Nonnullis ad Poema Pertinentibus*, defining it as the *study of attraction as concerned solely with discriminating perception*. For Baumgarten, in other words, aesthetics should be a separate, independent concern dealing only with perception. In Baumgarten's theory, much attention was concentrated on the creative act and the importance of feeling. For him, it was necessary to modify the traditional claim that "art imitates nature" by asserting that artists must deliberately alter nature by adding elements of feeling to perceived reality. In this way, the creative process of the world is mirrored in art activity. Baumgarten's thought in this regard was influenced by the philosophy of Gottfried Wilhelm Leibniz (1646-1716) and that of Leibniz's pupil Baron Christian von Wolff (1679-1754).

The problems of aesthetics had been treated by others before Baumgarten, but he both advanced the discussion of art and beauty and set the discipline off from the rest of philosophy. Immanuel Kant (1724-1804), who used Baumgarten's *Metaphysica* as a text for lecturing, retained Baumgarten's use of the term *aesthetics* as applying to the entire field of sensory knowledge. When combined with logic, aesthetics formed a larger discipline which Kant called *gnoseology*, a theory of knowledge that other philosophers called epistemology. Only later was the term aesthetics restricted to questions of beauty and of the nature of the fine arts.

Kant used the term *aesthetics* to argue that aesthetic appreciation reconciles the dualism of theory and practice in human nature, thereby leaving the way open to identify *beauty* (a relative, shifting and elusive concept) as a profoundly psychological quality (and not inherent in the artwork) by formulating a distinction between determinate, determinable, and indeterminate concepts. For him, beauty is non-determinate because we cannot know in advance whether something is beautiful or not by applying a set standard. He also deemed the concept of beauty non-determinable because, due to creativity, we will never find such a standard. Beauty, therefore, must reside in the indeterminate supersensible. And so must noise. This approach supports Walter Benjamin's (1892-1940) conception of aesthetics not as a part of a theory of the fine arts, but as a theory of perception.

For me, such an approach to noise aesthetics was anticipated by Georges Bataille (1897-1962)[33] when he considered excess the non-hypocritical human condition, which he took to be roused non-productive expenditure (excess) entangled with exhilaration. Excess, for Bataille, is not so much a surplus as *an effective passage beyond established limits*, an impulse which exceeds even its own threshold.[34] When one takes an interpretative metaphorical view of noise—broader than the typical, somewhat fatuous, reductive explanations—one soon detects that the concept of noise itself is an *open concept*.[35] The concept of noise itself is *pantheoristic*. But in my use of the term (based on my activities as an artist) I understand art noise to be fundamentally *an extravagant activity of creativity*.[36]

However, a new meaning for noise in art (and in life) is not developed by thinking about the aesthetics of noise as an invasive and unpleasant return to a dark primal unconscious.[37] It works when noise is also understood as an expanded[38] psychic thermidor: when it takes us back from the edge and rounds out our sensibilities as it forces us to get with the underlying assumptions of excess inherent in noise.[39] As Allen S. Weiss says in his incisive book *Phantasmic Radio*, "Noise creates new meaning both by interpreting the old meanings and by consequently unchanneling auditory perception and thus freeing the imagination".[40] This book is an attempt at facing up to the radical implications of those assumptions, and at purging us from conventional ways of thinking about noise (which is often disapprovingly).

This purging strategy provides a means of exemplifying various methods of thinking about noise. But it will not position noise in an easy opposition to Johann Joachim Winckelmann's (1717-1768) codification of classical ideals as those being primarily uncomplicated and Apollonian in their logic, for when the idea of simplicity takes on the intensity of a righteous injunction, the implied equation between simplicity and goodness obscures a less evident function, that of cognitive constraint.

Plus noise today has become too subjective[41] and ambient for that kind of dialectic between Apollonian calmness in relation to Dionysian non-restraint. And too sublime,[42] with its mix of alarm, approval, apprehension, and ascendancy. As Edmund Burke says in his *A Philosophical Enquiry into the Origin of Our Ideas of the Sublime and Beautiful*: "The passions which belong to self-preservation, turn on pain and danger; they are simply painful when their causes immediately affect us; they are delightful when we have an idea of pain and danger, without being actually in such circumstances; this delight I have not called pleasure, because it turns on pain, and because it is different enough from any idea of positive pleasure. Whatever excites this delight I call *sublime*". So this time, sublime immersion into noise promotes a conflicted but promiscuous ontological feeling (awareness/consciousness) where aesthetic cognition of the limits of the aesthetic attain the actual state of the generally subjective world of consciousness itself.

Pertinent to these noise concerns is, of course, Friedrich Nietzsche (1844-1900) and his acute criticism of the static culture of the bourgeoisie, particularly as it relates to the *gesamtkunstwerkkonzept* in *Die Geburt der Tragödie* (The Birth of Tragedy), Nietzsche's account of classical Greek drama and its merits. In *The Birth of Tragedy*, Nietzsche procures the concepts of the Apollonian and the Dionysian principles from Greek tragedy. According to Nietzsche, the Apollonian principle: reasoned, restrained, self-controlled and organizing is subsumed *within* the Dionysian principle: that which is primordial, passionate, chaotic, frenzied, chthonic and creative. This dialectical aesthetic tension allows the imaginative power of Dionysius to operate, in that the products of this operation are kept intelligible by Apollonian constraint.

Hence Nietzsche examined the dialectic between an Apollonian calmness in relation to an antecedent Dionysian non-restraining tragedy which has its origins in the chants of the Greek chorus. By invoking the power of the Greek drama, Nietzsche implied a pejorative judgement on subsequent dramatic forms of realism and inert spectatorship. Generally speaking, this aspect of Nietzsche's thought thus participated in the widespread ideal embedded in Romanticism[43] of a popular recovery of the mythic precondition necessary for cultural consciousness based on, in most cases, the sublime excess of the infinite.[44]

In ancient polytheistic Greece, sacred rites were in certain cases enacted on or near sacred grove sites. One such well-recorded rite was the ecstatic Dionysian rite. The Dionysian rite was directed not at the nymphs however, but to Dionysos (also known as Dionysius, Bacchus and/or Bakchos), the God of wine, intoxication and creative *ekstasis*. Dionysian ecstatic festivities were based on an even earlier form of ritual, the ancient Springtime Spree which was a three-day agricultural gala which involved the uncasking and drinking of that year's wine, the planting of seeds, and the encountering of ghosts.[45] By intoxicatingly mixing seeds with memories of their dead in the earth (which was viewed as the domain of the deceased) the ancient Greeks were able to incorporate their departed into the drinking and planting festival of the Spring Dionysia.

Subsequently the Spring Spree evolved into the even more intense rite of Dionysian *ekstasis* which intensified consciousness through drink and ecstatic prancing. The culmination of the Dionysian *ekstasis* rite was an ecstatic frenzy in which the dancers tore apart and devoured raw a sacrificial animal, such as a goat or a fawn. At the center of the rite are the mental states of *ek-stasis* and *en-thusiasmós*, states where psychological frontiers are torn down in preparation for the immersive divine dive into a world of animalistic unity.

As we will see further on, noise in art often evokes vast entropy[46] as a form of vacuole, even as entropy quantifies uncertainty in encounters with random variables. In his book, *Poétique de l'espace* (The Poetics of Space), Gaston Bachelard (1884-1962) speaks of the French poet Charles Baudelaire's (1821-1867) frequent use of the word *vast*, which is, Bachelard claims, one of the most Baudelairian of words: the word that marks naturally, for this poet, the infinity of the intimate. This, at first seemingly paradoxical statement, is correct, for Baudelaire is above all celebrated as a poet and practitioner of *double consciousness*: incarnating two intertwined natures. Apparent polar opposites play against (and ultimately with) each other dialectically in his thinking.

In comparable ways, art noise is both decadent and sublime when it is founded on vacuole principles of debauchment and self-indulgent consternation. Of course, such a dithyrambic logic has manifested in all modes of decadent artistic periods, from the Hellenistic and Flamboyant Gothic, to the Mannerist, Rococo, and Fin-de-Siècle, as they all opposed dogmatically imposed paradigms with hyper-engendering strategies. This is precisely what art noise does today. Because of noise's stimulation of the nerves, the decadent sublime (as safe terror) is agreeably engaging (particularly in grating harsh noise music)[47] even as this psychologically intense feeling is bound up with our sense of mortality in an existential way similar to the terms of Jean-Paul Sartre's (1904-1980) *Being and Nothingness*. Indeed the term *sublime* specifically refers to an aesthetic vacuole value in which the primary factor is the presence or suggestion of undivided vastness and immense breadth of space, which is incapable of being completely ascertained.

Noise art (musical and visual) therefore never offers us conventions. Rather, when good, it is like a fertile seedbed which undermines the hitherto clear, false distinctions between representation (identity) and the imagination by way of negating and recombining. Here, semblances and sounds are always already connected within a crushed and dark and obscure excessive orb, as noise art negates artistic representations (and all they imply), thereby affirming a consciously divergent way of perceiving and existing. Such excessive artistic noise can therefore spawn in us a sense of affinity which communicates individuality in totality without forfeiting liberty.

I should say that several of my ideas on this subject of noise stemmed from reading Georges Bataille's *Visions of Excess* (which appeared in English translation in 1985) after which I began to experiment with (and analyze through my artwork) various artistic approaches towards noise and excess.[48] In the terms Bataille proposes, any "restricted economy", any sealed arrangement (such as an image, an identity, a concept or a structure) produces more than it can account for, hence it will be inevitably fractured by its own unacknowledged excess and, in seeking to maintain itself, will, against its own rationalized logic, crave rupture, expenditure, and loss. More specifically, for Bataille, the term *expenditure* describes an aspect of erotic activity poised against an economy of production.[49] Yet Bataille's accomplishment transgresses disciplines and genres so repeatedly and so thoroughly that capsule accounts of his work in terms of noise are compelled to delegate themselves to abstractions. One can say with assurance that his thinking consisted of a meditation on, and fulfil-

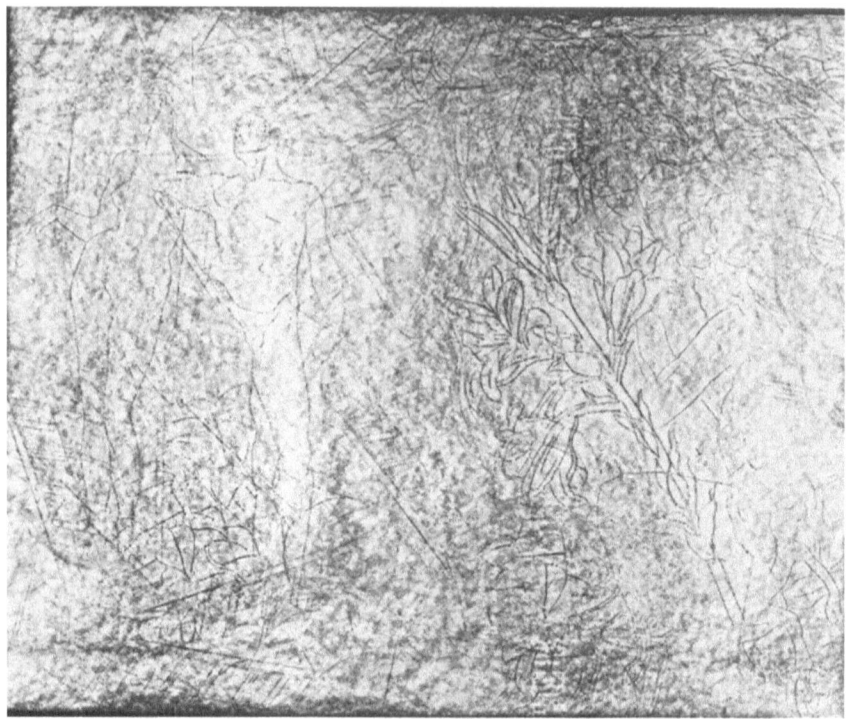

Figure 2 – *Palace of Power*, 11x14" graphite on paper, 1984, Joseph Nechvatal

ment of, *transgressions through excess*. Thus Bataille's *Visions of Excess* immediately impressed me as it resonated handsomely with the overloaded nature of my palimpsest-like grey graphite drawings from the early-1980s (which were reflective of the time's concerns with the proliferation of ideology connected to the proliferation of nuclear weapons).

So via Bataille, I can say that noise art's probing at the outer limits of recognizable representation, the excited all-over fullness and fervor of this syncretistic probe, isn't a failing of communications in the art; *it is its subject*. Good noise art, for me at least, is capable of nurturing a sense of polysemic uniqueness and of individuality brought about through its *counter-mannerist* style (circuitous, excessive and decadent); it is a style that takes me from the state of the social to the state of the secret, distinguishable I by overloading ideological representation to a point where it becomes non-representational. It is this *non-representational counter-mannerist representation* which breaks us out of the fascination and complicity with the mass media mode of communication. Noise art frees us, then, from accustomed coyness, platitudes, and predetermined perceptions with which we are deluged daily by the mass-pop media. It is my experience that it is in this artistic condition of privately excessive formlessness that we can ascertain the delimitation of mass-pop media ideology and the resultant implications of that cognizance.

Noisy Methodology

The method used here reflects on the insights noise suggests to the traditional western history of unified being (which indeed engenders extraordinarily deep conflicts). This will entail a review of past and present approaches towards ontology and an analysis of a variety of artistic maneuvers. I will non-teleologically synthesize these questions and examples of ontology into an interrelated theoretical model for noise art by clarifying its underlying philosophy of significance. I will thus outline an integrative noise philosophy by tracing the visual noise impetus through its various expressions so as to examine the immersive noise philosophy from all possible sides. Of principal interest will be the discussion of *subject/object* cognition.

Even though Otto Kernberg pointed out that the splitting of the subject from the object is "the crucial mechanism for the defensive organization of the ego" at its most basic (pre-oedipal) level,[50] the subject/object question pursued in this discussion will not appear in any stable binary positioning[51] of easy subject/object opposites as I recognize, as Stephen Talbott points out, that the subject/object set functions more along the dialectical lines of the magnet, where the north pole exists only by virtue of the south pole (as is the contrary). Like the supposed subject/object opposites, neither pole exists in isolation.[52] Hence a subject/object debate in terms of immersive perspective (a debate I do not wish to shy away from) is possible only with the radical conflation of this polarity into an *omnijectivity* (see below) which recognizes the mutual interpenetration that unites the apparent opposites. Then there is something of the subject in the most impenetrable object, and an objective, world-sustaining presence in the sheerest subject. As with the magnet, where if you nip off the slightest piece of one end of the magnet you will discover that it still possess both a south pole and a north pole, so the forces of subjectivity and objectivity co-exist in omnijectivity. It is as impossible to conceive of an isolated subject or an isolated object as it is to conceive an isolated north or south pole, but it is entirely imaginable to relinquish sight of their conjoint importance.

Useful here is the concept of *omnijectivity* (the metaphysical concept stemming from the discoveries of quantum physics which teaches us that mind (previously considered the subjective realm) and matter (previously considered as the objective realm) are inextricably linked in reconciling the relativist ("subjective") style found here mixed with the more absolutist ("objective") style—as omnijectivity is possible only with the conflation of polarities. It is a stance which recognizes the mutual interpenetration that unites the apparent opposites of subjectivity and objectivity. More specifically, the concept of omnijectivity emerged from the theories of quantum physicist David Bohm (1917-1992)—protégé of Albert Einstein (1879-1955)—and Karl Pribram, author of the neuropsychological textbook *Languages of the Brain*. Pribram noted that modern theories of how the brain stores memories did not explain how memories seem to be distributed throughout the brain as a whole. Each memory a person has was believed to have a specific location somewhere in the brain cells. Pribram, however, made the discovery that memories are not localized, but are somehow spread out or distributed throughout the brain as a whole. Even when considerable damage is done to a brain, or pieces of it are removed, organisms don't lose sections of their memory. He knew of no process that could account for such a phenomenon. Finally, the process that made the most sense in metaphorically explaining this aspect of the brain was *holography*. Therefore Pribram offered the holographic model as an explanation of the functioning of the brain.

The term, *omnijectivity*, corresponds with Gene Youngblood's term *extra-objective*, which he used to describe the synaesthetic and psychedelic features of what he termed *synaesthetic cinema*, an underground cinema tendency of the late-1960s that ostensibly combined subjective, objective, and non-objective features into a syncretistic perception of the simultaneous space-time continuum.[53] This syncretistic perception was chiefly accomplished by the use of superimposition and by "reducing depth-of-field to a total field of non-focused multiplicity" after closing the span between the inside and the outside of the picture plane.[54] Youngblood derived the term *synaesthetic* from Anton Ehrenzweig's (1908-1966) idea of *syncretistic vision*, which Ehrenzweig characterized in his book, *The Hidden Order of Art* as a *Total Vision*.[55]

An immersive noise aesthetic beckons and amends the *mind/body problem*, the metaphysical problem of how the mind and body (and I would stress the body's ears and eyes) are related to one another, and of how consciousness relates to conjectural substantiality in immersion. Aesthetic immersive consciousness, particularly when comprehended as noisy, may be said to be in a vibratory self/non-self referential mode and thus illustrative of what Metzinger sees as the "infinitely close and at the same time infinitely distant"[56] characteristic implicit in all states of consciousness. This pre-pleonastic vibratory comprehension, which illumes Metzinger's attestation, occurs by way of the distance that the artifice of immersive noise art confers to consciousness—an artifice which lends itself to a reactive, self-attentive unification. As such, immersive artifice works to circumvent the current fragmentary view of the body/mind in the world which has been underpinned by the Cartesian/Newtonian model of optical physics.

Based on these understandings of immersive consciousness, the abstract, immersive noise theory proposed here could develop a means for achieving insight into how the agency of aural and visual thought works when we release them from their *compos mentis* obligations.[57] To achieve such an examination would be to overcome the tendency for aesthetic visual thought to analyze itself in terms of a presumed separation between the process of visual thinking and the content of visual thought which is its product, and this view of immersive consciousness clarifies an initial issue of immersion into noise in one grand sense. Since visual thinking is shown to be a process consisting of the transformations of nuero-physical visual-thought impulses impregnated in continuous waves, our visual thoughts are not distinct from visual thinking. Similarly, immersive visual thought, visual thinking, and visual thinker make up a reverberating, incessant, multiple, and unified continuity.

I realize that this comprehension is nothing more than visual aesthetics catching up to basic science today. As the analytical philosopher Thomas Metzinger says, "in the physical outside world there are only electro-magnetic oscillations of certain wavelengths", and that in a scientific look at reality, "all we find are myriads of subtle electrical impulses".[58] Taking it a step further in seeking the field of contact between the inner cognitive world and the outer penetrable world of physics (a realm which I posit is the veritable domain of art), it makes sense to see thinking, thought, self, and experienced immersion into noise as a non-localized flow of reverberating, incessant, multiple, but hyper-unified frequencies in which self-conscious manifestations occur through an awareness of the richness of noise.

Among the more important positions in the formation of this general debate is René Descartes' (1596-1650) argument that the mind and body are quite disconnected elements that nevertheless interact with one another. Richard Rorty (1931-2007) asserts that Descartes' feat was to conceive of the human mind as an internal chasm in which both pains and clear and distinct ideas passed in examination before an inner eye. By contrast, Gottfried Wilhelm Leibniz asserted a theory of psycho-physical parallelism based on his theory of *Monadology*, his model of a system which conceived of unity in plurality and plurality in unity. Leibniz's monadological ideas have substantially influenced Gilles Deleuze, whose fertile philosophical articulations have played an important role for me.

Based on the above understanding of aspects of qualified hyper-absolutes and post-modern omnijective understandings concerning immersion into noise, my formative contention is that immersive aesthetics—when re-contextualized in a wider historical arena—can be reasonably adept in assisting us in the intrinsically hermeneutical comprehension of our existence. Such an omnijective/immersive aesthetic could be capable of heightening the relative theoretical worth of art historical scholarship in relation to the most recent developments of the information revolution, in the service of an expansive conversation concerning our aesthetic self-consciousness.

But for us to get started on the road to a fuller aesthetic self-consciousness, I wish to assert the idea of *hyper-noise*[59]—a notion of *noise as art constructed via connected-competing vectors*.[60] This anti-pure approach suggests an encircling mental ideal[61] that allows unaccustomed creative situations and sensations to connect and tolerantly co-exist. This hyper-noise idea is based on my understanding of the flat random aspects of white noise in conjunction with the all inclusiveness inherent in white light. In hyper-noise, signal-to-noise touch, run over and become akin to each other, so that self-centered human comprehension may both deepen and spread into an expansive constitution of being. This being as centered vitality might be described, as John Cage (1912-1992) suggests, as "central to a sphere without surface, [...] unimpeded, energetically broadcast [...] as transmission in all directions from the field's center".[62]

So art as white-noise-light speaks to us both at the center and at the edges of our frame of cognition (that frame semi-forced on us by the social and psychological conditioning of empiricist/positivist philosophy). And, as such, white noise art might allow us to feel the outer limits and finer levels of our most exquisite vacuole sensibilities.

Because we will never succeed completely in understanding these feelings, white noise art is tragic. Because we never stop trying, white noise art is comic.

Towards a Noise Vision: the revolution will be visualized

Following a short discussion of *Noise Music* (I point the reader to Paul Hegarty's book *Noise/Music,* Caleb Kelly's *Cracked Media: The Sound of Malfunction,* Brandon LaBelle's *Background Noise: Perspectives on Sound Art,* Salome Voegelin's *Listening to Noise and Silence,* Thomas Bey William Bailey's *Micro Bionic: Radical Electronic Music And Sound Art In The 21st Century,* Joanna Demers's *Listening Through The Noise* and Steve Goodman's *Sonic Warfare* for a fuller audio/music-specific discussions of noise) I will exemplify my notion of visual hyper-noise with the physically contained (but optically boundless) palimpsest-esque, all-over, wall-paper-like image spread found in the Apse of Lascaux, that I experienced. Here I will speak of a kind of *eye noise*: a distinct visual-cognitive proclivity that addresses the multiplicitous/heterogeneous impetus within a visual aggregate. I will define this *visual noise* as being produced by an all-over, elaborate, spread out distribution of visual incident which calls upon the optic procedure of spatial summation—a process which unconsciously totalizes the visual excess encountered. I will refer to this *noise aggregate* as a *summational but all-over net-condition/awareness of plurality in hyper-homogeneity, a supplementary order of diversity within orders of noise*. Such creative noise, I will suggest, is capable of creating new forms of order. This creative condition will relate to what I will call *hypercognitive noise* with respect to my conclusions, conclusions that recognize that from now on things will be heard and seen only from the connected depths of a noisy and nervous inclusive density, withdrawn into itself, perhaps, adumbrated and darkened by its obscurity, but bound tightly together and inescapably grouped by the vacuole vigor that is noise.[63] I submit that such conclusions are capable of fermenting a phantasmagorical discourse that is both nervously capricious and, perhaps paradoxically, socially responsible.

Accordingly, I wish to present the idea of noise art as that *art which precludes established significance by replacing the assumption of conclusive meaning with one of vital excess* (a non-objective communication of emotional significance). A noisy hyper-cognitive is where the particular of electronic connectivity is seen as part of an accrual total system by virtue of its being connected to everything else while remaining dissonant.

The strategy of a dissonant hyper-anything includes principles of networked connections and electronic links which give multiple choices of passages to follow and continually new branching possibilities. The total-hyper-being model for a new connected noise noology[64] (especially when placing emphasis on tabulating an evasive orb) is the self-re-programmable internal function which explicitly offers a furtherance in envisioning internal, anti-hierarchical models of thought to ourselves. As such it is a procedure of epistemological symbolization.[65]

This self-connected epistemological-strategic approach is based on the premise that behind all noise art, either representational or abstract, is the hypothetical exploration of the introspective rhizomatizing world of the imagination under the influence of today's high-frequency, electronic/computerized environment. Any analogue-to-digital conversion process transfigures various physical quantities into homogeneous numbers. And numbers, it must be remembered, are abstractions that have no solid tangible actuality. Moreover, since it is difficult to make sense of today's swirling, phantasmagorical media society, the general proposition behind noise art may best be to look for a paradoxical summation of this uncertainty by taking advantage of today's superficial image saturation—a saturation so dense that it fails to communicate anything particular at all upon which we can concur, except perhaps its overall incomprehensible sense of ripe delirium as the reproduction system pulses with higher and higher, faster and faster flows of digital data to the point of near hysteria.

Perhaps the result of this ripe information abundance is that the greater the amount of information that flows, the greater the non-teleological uncertainty that is produced. Hence noise. So, the tremendous load of imagery/sound/text information digitally produced and reproduced all round us today ultimately seems to make less, not more, conventional teleological sense. Information as knowledge is myth.

This seems to be in agreement with Sigmund Freud (1856-1939), as he too thought that the boundaries which make up the various territories of art historical knowledge needed to be transcended by their relation to the depth of mythic representation.[66] As a result, in my view, it is noise art's obligation to see what mythic, unconventional, paradoxical, summational sense—in terms of the rhizomatizing world of the imagination—noise might make of all this based on an appropriately decadent reading of our paradoxically material-based (yet electronically activated) social media environment.

This formulation is familiar with respect to the fantastical aspects projected into the qualities of an idealized situation (as we know from many kinds of antediluvian religious metaphysics and their artistic transcendental expressions). Salient here is that the function of the symbol is to (supposedly) intellectually transposition people momentarily to other realms of reality. Indeed one prime aspect of any noise is its ability to relocate consciousness into another aspect of the world.[67] Noise art, then, is a form of deviant encoding that precipitates inner shifts within a communicative space by use of an annoying grammar that can smash together and interfere with neighboring discourses.

This subject of noise vision, and the rhetorical strategy needed to explore it, especially interests me in that encounters with noise art (one may assume) could create an opportunity for personal transgression and for a vertiginous ecstasy of thought.[68] As I suggested before, noise-based perception-cognition (awareness) requires a plenum consciousness where there is only the slightest difference between an intentional and an involuntary exceeding of representation. Such an explosive collapse of utilitarian consciousness (combined with the pursuit of inexactitude), I wish to suggest, may fashion an abstract intensity within our perceptual circuitry—hence exceeding the assumed determinism of the technological-based phenomenon inherent (supposedly) in our post-industrial information society.

The history of art is, of course, full of new epistemological shifts and I maintain here that the shift in perspective which noise provides is just such a shift, replete with a newness based on a long preparatory gestational development.[69] Indeed, it seems to me that as human psychic energies are stifled and/or bypassed by certain controlling aspects of mass informational technology, such a personally transgressive ecstatic phenomena will most likely increasingly break out in forms of noisy thinking—resulting in noisy art.

In closing this introduction, I just want to mention that in my view one of the most important characteristics of immersion into noise is its sense of encompassing being within a field of vibratory enshrouding: an intertwined embossed shrouding that places us at odds with the closed, cliché, visual-audio signal and resituates us within vibrancy.

The emergence of the theoretical project I have outlined above will, I hope, contribute to the surpassing of thought representations by inventing a noisy thinking/art in which what matters is no longer only uninterrupted identities, or logos, or distinctive characters,[70] but rather a lush, phantasmagorical clamor developed on the basis of inclusion.[71] Such dynamic abstract thinking (and the art forms that result) I hope will be presented to us only in an already pre-connected vivacious state of noise, already articulated in that insinuated clatter that is linking us in a vacuole discourse which is both non-teleologically oriented and intellectually responsible.

CHAPTER 1

Into Noise: Tabula Rasa vs. Horror Vacui

I consider myself successful only when I do something that resembles the lack of order I sense.
— Robert Rauschenberg

Is not perhaps all ecstasy in one world humiliating sobriety in that complementary to it?
— Walter Benjamin, *Reflections*

To set in motion the above reflection, I will address my experiences and understandings of noise as an immersive recalcitrance cultural characteristic through its sonic aspects.[72] That is how noise is generally understood. But then (perhaps more perplexingly) I will address noise as constructive visual, conceptual and theoretical phenomenon that has both communicative and philosophical ramifications connected to the unsound and the intricate.

Noise is a dark and thorny thought, for the notion of noise does not have an unchanging, artistic unworthiness or worth. Rather noise is what lies outside of our habitual comfort zone at any point in time.[73] It is what awakes us from our silent dream and leads us to the excesses of ecstatic encounters.[74] It is non-homogeneous and incomplete, while being always hermetic. At the same time, noise is predominantly phantasmagorical. It suggests an outside other and points us elsewhere by sabotaging our sense of harmonious balance. Hence it is a corrective to efficiency and aesthetic correctness and an urge for psychic newness, as we feel compelled to move on. In this sense it is aligned with a Dada-like illogical determination.[75]

One consequential effect of noise is on connected-internal cognizant models, such as the psychological ego-center and the subject's psychological motivational drive; these are factors which raise felt intensity.

Hence the persuasiveness of noise. This is demonstrated by the fact that one enjoys noise as art (or doesn't) depending to a large extent on one's personal psychological needs and adaptability in accordance with the proposed audio cues. The philosophic rhizomatic theory of Gilles Deleuze and Félix Guattari (1930-1992), at a general level, supports such a connectivist approach towards theorizing the noise experience, as rhizomatic theory encourages philosophic non-linear and non-restrictive interdisciplinary thinking and hence reinterpretation, which in this case will proceed from the point of view (not a point of fact anymore, but an orb) of *immersion into noise*. A *rhizome* literally is a root-like plant stem that forms a large entwined spherical zone of small roots that criss-cross. In the philosophical writings of Deleuze and Guattari, the term is used as a metaphor for an epistemology[76] that spreads in all directions simultaneously.[77] More specifically, Deleuze and Guattari define the rhizome as that which is "reducible to neither the One or the multiple. [...] It has neither beginning nor end, but always a middle (*milieu*) from which it grows and which it overspills. It constitutes linear multiplicities with n dimensions having neither subject nor object".[78]

Noise music often delivers a rhizomatic Dionysian jittery quality packed with levels of complexity, if not always indulgent Dionysian chaos. Certainly, noise (as art) merits the adjective *polysemic*, a word which stems from the Greek phrase meaning *many signs*. A polysemic awareness of noise acknowledges the hypothetically infinite range of meanings of noise that result when determinacy is replaced by indeterminacy, an awareness which contradicts the verisimilitude thought to correspond to the assumed exactitude of naive naturalism. Noise art is properly concerned with *ideals of self-transcendence*, then.

Nevertheless, for the sake of discussion, the reader will, I hope, consider my somewhat less precise and less formal definition of noise art as *boundary-quivering creation that involves the destruction of the conceptual frame* that often uses temporary excess coupled with negativity (in some sense). In noise art, even a form of negation like *anti-art*[79] functions finally as a way of opening up our capacity for plurality and expanding what has heretofore been accepted.[80] The preferred decisive point in understanding immersion into noise in the context of art is its facilitation of a more potent *conscious-totality* in the art audience produced by merging the audience's perceptual circuitry with the artwork.

Ultimately for me, though, noise is just a *rupture signifying transmission of excess and/or negativity* for the artist to employ or disregard at will. It can be lavish and thrilling. It can be incredibly tedious and boring.

Good noise music has the proclivity to solicit us to respond to the work in time in both a contemplative and physical way, or at least in an implied tension between these two poles when one side outweighs the other. Noise, as all sound, dissipates over time, hence the only consistent non-expansive definition of the term *noise* that works for me in the context of the arts is in its *irreversibility of time*.

As it is bound to dissipate over time, noise is death hiding in life, and it is true that expectations of death clearly condition our sense of boundaries. Our prospects for an everlasting life are not so drastically different than they have ever been. So perhaps beauty and noise don't care for each other much. Nevertheless, in respect to noise, let's consider Kant's formulation of his *sublime ideal*—something that moved philosophical inquiry away from what was weakly established as the objective ideal in which the world, and the human subject within it, could be described as if from an outside (logocentric) position.[81] Kant, in *Observations on the Feeling of the Beautiful and Sublime*, showed us that intellectually humans are incapable of knowing the sublime ultimate reality, but this need not, and must not, according to Kant, interfere with the human obligation of performing as though the ideal character of this reality were certain.

So without being certain, I will embark on this inharmonious meditation of noise discernment by remembering how I first became aware of noise as a cultural agent with aesthetic power.

Tabulating Cybernetic Hendrix

My childhood was set in a rather peaceful suburban setting. The distant sound of the train is the only noise I can recall, other than occasional thunder and other sounds of nature. My first cultural aspect of noise occurred at age 17 when I attended *The Jimi Hendrix Experience* concert December 1, 1968 at the Chicago Coliseum and sat in the very last row—far far away from the stage. Hendrix appeared miniscule. However the speakers were located just behind my head and his grandiloquent feedback was ear-splitting; an intensely pleasant, if disjunctive noise experience.

That day I felt and sensed that this fabulous feedback I had been hearing was expressing inner aspirations for my own personal cybernetic circumstances, exemplified by Hendrix's ongoing mixing of the electronic with the loose gesture. As pointed out by John Johnston in his *The Allure of Machinic Life: Cybernetics, Artificial Life, and the New AI*, a special feature of cybernetic theories (theories of feedback systems primarily based on the ideas of Norbert Wiener, 1894-1964) is that they explain processes in terms of the organization of the system manifesting it (*e.g.*, the circular causality of feedback-loops which enables cybernetics to elucidate complex relationships from within—useful in formulating a creative epistemology concerned with the self-communication within a psyche and between the psyche and the surrounding environment).[82] I believe that via my confrontation with the exterior spectacle of Jimi's noise music, an intimate grandeur unfolded for me connected to noise for the first time and a private readiness to amalgamate dislocated profusion was quickened. Indeed in reading Johnston's *The Allure of Machinic Life*, I came across a quotation by Wiener on cybernetics[83] where he discusses the "study of automata", pointing out that one of its "cardinal notions" is the "amount of disturbance or 'noise'" engaged.

Of course, Hendrix creatively used self-generating noises—the sounds that used to be denounced as non-musical—therefore it seemed to me, he understood that through the mediation of machines the technological built-in can be contorted and bent, thus changing our awareness of what technology is or can be. In that sense, his use of noise as a musical element demonstrated to me for the first time how an art of superabundance-proliferation can also be an art of pleasure that enlivens us to the privateness—and unique separateness—of each of us in lieu of the constructed social spectacle that engulfs and (supposedly) controls us (through technology). This private excess/noise created a separateness that offered me a personal critical distance via a bacchanalia gap with which to problematize technology—and thus another perspective on (and from) my given social simplicity. Thus noise as music is rhizomatic and increases our cultural territory by moving towards deterritorialization. As such, noise music demands a different kind of hearing and feeling not always belligerent and thunderous.

Noise Music

Generally speaking, *Noise Music* is a term used to describe varieties of avant-garde music and sound art that may use elements such as cacophony, dissonance, atonality, noise, indeterminacy, and repetition in their realization. In defining noise music and its value, Paul Hegarty cites the work of noted cultural critics Jean Baudrillard (1929-2007), Georges Bataille and Theodor Adorno (1903-1969) and, through their work, traces the history of noise. He defines noise at different times as "intrusive, unwanted", "lacking skill, not being appropriate" and "a threatening emptiness", and he traces these trends starting with 18th century concert hall music. Hegarty contends that it is John Cage's composition 4'33"—in which an audience sits through four and a half minutes of "silence"—that represents the beginning of noise music proper.[84] For Hegarty, *noise music*, as with 4'33", is that music made up of incidental sounds that represent perfectly the tension between "desirable" sound (properly played musical notes) and undesirable "noise" that make up all noise music from Erik Satie (1866-1925) to NON to Glenn Branca.

Douglas Kahn, in his work, *Noise, Water, Meat: A History of Sound in the Arts*, discusses the use of noise as a medium and explores the ideas of Antonin Artaud, George Brecht, William Burroughs, Sergei Eisenstein, Fluxus, Allan Kaprow, Michael McClure, Yoko Ono, Jackson Pollock, Luigi Russolo and Dziga Vertov.[85]

In *Noise: The Political Economy of Music*, Jacques Attali explores the relationship between noise music and the future of society.[86] He indicates that noise music is a predictor of social change and demonstrates how noise acts as the subconscious of society, validating and testing new social and political realities.[87]

Like much of modern and contemporary art, noise music takes characteristics of the perceived negative traits of noises and uses them in aesthetic and imaginative ways. One can find the distinct effort to create something harshly beautiful from something perceived as ugly in what can possibly be identified as a search for a post-industrial sublime in art.

In much the same way that the early modernists were inspired by naïve art, some contemporary digital art noise musicians are excited by the archaic audio technologies such as wire-recorders, the 8-track cartridge, and vinyl records. Many artists not only build their own noise-generating devices, but even their own specialized recording equipment and custom software (for example, the C++ software used in creating my *viral symphOny*).[88]

For me, art noise combines stimulation into an all-inclusive totalization through *sympathetic vibration*, just as strings of a piano vibrate in sympathetic agreement, especially when tuned to the tuning system called *just intonation*. Just intonation, in music, is a system of tuning in which the correct size of all the intervals of the scale is calculated by different additions and subtractions of pure natural thirds and fifths (the intervals that occur between the fourth and fifth, and second and third tones respectively, of the natural harmonic series). Supposedly used in medieval monophonic music (melody without harmony) and considerably discussed by 20th-century sound artists and art-music theorists, just intonation proved impractical for polyphonic (multi-part) music and was replaced at least by the year 1500 by meantone temperament.

Noise art music can feature distortion,[89] various types of acoustically or electronically generated noise, randomly produced electronic signals and non-traditional musical instruments. Noise music may also incorporate manipulated recordings, static, hiss and hum, feedback, live machine sounds, custom noise software, circuit bent instruments, and non-musical vocal elements that push noise towards the ecstatic. The Futurist art movement was important for the development of the noise aesthetic,[90] as was the Dada art movement[91] (a prime example being the *Antisymphony* of Jefym Golyscheff performed by Hannah Höch in Berlin on April 30th, 1919 with kitchen utensils)[92] —and later the Surrealist and Fluxus art movements, specifically the Fluxus artists Joe Jones, Yasunao Tone, George Brecht, Wolf Vostell, Yoko Ono, Walter De Maria's *Ocean Music*, La Monte Young, Robert Watts,[93] Takehisa Kosugi and Milan Knizak's *Broken Music*.[94]

During the early 1900s, a number of art music practitioners began exploring atonality. Composers such as Arnold Schoenberg proposed the incorporation of harmonic systems that were, at the time, considered dissonant. This guided the development of twelve-tone technique and serialism. In *The Emancipation of Dissonance*, Thomas J. Harrison, in 1910, suggested that this development might be described as a metanarrative to justify the so-called Dionysian pleasures of atonal noise.[95] Contemporary noise music is often associated with excessive volume and distortion, particularly in the popular music domain with examples such as Boys Noize, Jimi Hendrix's previously mentioned use of feedback, Nine Inch Nails and Lou Reed's *Metal Machine Music*.

Other examples of music that contain noise-based features include works by Iannis Xenakis, Karlheinz Stockhausen, Helmut Lachenmann, Theater of Eternal Music, Cornelius Cardew, Rhys Chatham, Ryoji Ikeda, Survival Research Laboratories, Whitehouse, Cabaret Voltaire, Psychic TV, Jean Tinguely's recordings of his sound sculpture (specifically *Bascule VII'*), the music of Hermann Nitsch's Orgien Mysterien Theater, and La Monte Young's bowed gong works from the late 1960s, for example *23 VIII 64 2:50:45—3:11 am The Volga Delta From Studies In The Bowed Disc'* from *The Black Record* (1969). Genres such as industrial, industrial techno, and glitch music exploit noise-based materials.

Luigi Russolo, futurist painter of the very early 20th century, was perhaps the first noise music artist.[96] His 1913 manifesto, *L'Arte dei Rumori*, translated as *The Art of Noises*, stated that the industrial revolution had given modern men a greater capacity to appreciate more complex sounds. Russolo found traditional melodic music confining and envisioned noise music as its future replacement. He designed and constructed a number of noise-generating devices called Intonarumori and assembled a noise orchestra to perform with them. A performance of his *Gran Concerto Futuristico* (1917) was met with strong disapproval and violence from the audience, as Russolo himself had predicted. None of his intoning devices have survived, though recently some have been reconstructed and used in performances. Although Russolo's works have little resemblance to modern noise music, his pioneering creations cannot be overlooked as an essential stage in the evolution of this genre, and many artists are now familiar with his manifesto.[97]

An early Dada-related work from 1916 by Marcel Duchamp also worked with noise, but in an almost silent way.[98] His ready-made, *With Hidden Noise (À bruit secret)*, was a collaborative exercise that created a noise instrument that Duchamp accomplished with Walter Arensberg. What rattles inside when *With Hidden Noise* is shaken remains a mystery. By the 1920s, modernists Edgard Varèse and George Antheil began to use early mechanical musical instruments—such as the player piano and the siren—to create music that mirrored the noise of the modern world. Antheil's best-known noise composition is his 30 minutes-long *Ballet mécanique* (1924), originally conceived as the musical accompaniment to the Dada film of the same name by Dudley Murphy and Fernand Léger. Eventually the filmmakers and composers chose to let their creations evolve separately, although the film credits still included Antheil. Nevertheless, *Ballet mécanique* premiered as concert music in Paris in 1926. In 1920, the poem *Bruits* (*Noises*) was created by Lajos Kassák, a Hungarian painter, publisher and writer who made several visual poems with newspaper fragments and the superimposition of letters and graphics. In this poem, noises are interpreted as interferences of the media.

Antonio Russolo, the brother of the more famous Luigi Russolo, was another Italian Futurist composer. A 78rpm record made by him in 1921 is the only surviving sound recording that features the original intonarumori. Both pieces, *Corale* and *Serenata*, combined conventional orchestral music set against the famous noise machines. Filippo Tommaso Marinetti also assembled noises into a collage in which silence is an integral part. In 1923, Arthur Honegger created *Pacific 231*, a modernist musical composition that imitates the sound of a steam locomotive. Arseny Avraamov's composition *Symphony of Factory Sirens* involved navy ship sirens and whistles, bus and car horns, factory sirens, cannons, foghorns, artillery guns, machine guns, hydro-airplanes, a specially designed steam-whistle machine creating noisy renderings of *Internationale* and *Marseillaise* for a piece conducted by a team using flags and pistols when performed in the city of Baku in 1922.

In 1930, Paul Hindemith and Ernst Toch recycled records to create sound montages and in 1936 Edgard Varese experimented with records by playing them backwards and varying the playback speeds. John Cage started his *Imaginary Landscape* series in 1939, which combined recorded sound, percussion, and, in the case of *Imaginary Landscape #4*, twelve radios.[99]

In the 1940s, Pierre Boulez (who made his name with violently expressive scores and opinionated polemics) embodied a strict sound style shorn of Romantic nostalgia and the detritus of a defunct tradition.[100] Boulez moved on to the rigorously organized technique of total serialism, which organized various aspects of sound—pitch, duration, volume, and attack—into a series of twelve, in line with the twelve-tone system. Under the influence of Henry Cowell in San Francisco, Lou Harrison and John Cage began composing music for "junk" percussion ensembles, scouring junkyards and Chinatown antique shops for appropriately-tuned brake drums, flower pots, gongs, and more.

In Europe, during the late 1940s, Pierre Schaeffer coined the term *musique concrète* to refer to the peculiar nature of sounds on tape, separated from the source that generated them initially. Following this, both in Europe and America, other modernist art music composers such as Karlheinz Stockhausen, G.M. Koenig, Pierre Henry, Iannis Xenakis, La Monte Young, and David Tudor, explored sound-based composition. In late 1947, Antonin Artaud (1896-1948) recorded *Pour en Finir avec le Jugement de dieu* (*To Have Done With the Judgment of God*), an audio piece full of the seemingly random cacophony of xylophonic sounds mixed with various percussive elements, mixed with the noise of alarming human cries, screams, grunts, onomatopoeia, and glossolalia.[101]

In 1949, Nouveau Realisme artist, Yves Klein, wrote *The Monotone Symphony*, a symphony that consisted of one held note, thereby demonstrating that the sound of one sustained tone made viable music. Also in 1949, Boulez befriended John Cage, who was visiting Paris to do research on the music of Erik Satie. John Cage had been pushing music in even more startling directions during the war years, writing for prepared piano, junkyard percussion, and electronic gadgetry. In Paris, Cage encountered the pioneering electronic composer Pierre Schaeffer, who, after the war, began assembling sound collages made up of pre-recorded pieces of tape. The first of Schaeffer's *Cinq études de bruits*, or *Five Noise Etudes*, consists of locomotive sounds that the composer recorded at a train station.

Back in New York in 1952, Cage constructed his own tape collage, *Williams Mix*, made up of some 600 tape fragments arranged according to the demands of the *I Ching*. Cage's early radical phase reached its height that summer of 1952, when he unveiled the first art *Happening* at Black Mountain College, and 4'33", the so-called controversial *silent piece*. The audience saw David Tudor sit at the piano, and close the lid. Some time later, without having played any notes, he opened the lid. A while after that, again having played nothing, he closed the lid. And after a period of time, he opened the lid once more and rose from the piano. The piece had passed without a note being played, in fact without Tudor or anyone else on stage having made any deliberate sound, although he timed the lengths on a stopwatch while turning the pages of the score. Only then could the audience recognize what Cage insisted upon: that there is no such thing as silence. Noise that may make musical sound is always happening.

In 1957, Edgard Varèse created on tape an extended piece of music using noises not usually considered "musical" entitled *Poème électronique*. Varèse conceptualized his last work in the immersive, conceiving his unfinished *Espace* as voices in the sky, as though magic, filling all space, criss-crossing, overlapping, penetrating each other.

Among the techniques used in this period were tape manipulation, subtractive synthesis, and improvised live electronics.

On May 8th, 1960, six young Japanese musicians, including Takehisa Kosugi and Yasunao Tone, formed the Group Ongaku with two tape recordings of noise music: *Automatism* and *Object*. These recordings made use of a mixture of traditional musical instruments along with a vacuum cleaner, a radio, an oil drum, a doll, and a set of dishes. Moreover, the speed of the tape recording was manipulated, further distorting the sounds being recorded.

The art critic Rosalind Krauss argued that, by 1968, artists such as Robert Morris, Robert Smithson and Richard Serra had entered a situation the logical conditions of which can no longer be described as modernist. Sound art found itself in the same condition, but with an added emphasis on distribution. Anti-form process art became the term used to describe this post-modern, post-industrial culture and the process by which it is made. Serious art music responded to this conjuncture in terms of intense noise, for example the La Monte Young Fluxus composition *89 VI 8 C. 1:42-1:52 AM Paris Encore From Poem For Chairs, Tables, Benches, Etc.* Young's composition *Two Sounds* (1960) was composed for amplified percussion and windowpanes, and his *Poem for Tables, Chairs and Benches* (1960) used the sounds of furniture scraping across the floor.

In addition, a process of anti-form *free noise* emerged out of the avant-garde jazz tradition with musicians such as John Coltrane, Pharoah Sanders, Ornette Coleman, Cecil Taylor, Eric Dolphy, Archie Shepp, Sun Ra and the Arkestra, Albert Ayler, Peter Brötzmann, and John Zorn. In the 1970s, the concept of art itself expanded and groups like Survival Research Laboratories, Borbetomagus and Elliott Sharp embraced and extended the most dissonant and least approachable aspects of these musical/spatial concepts.

Around the same time, the first postmodern wave of industrial noise music appeared with Throbbing Gristle, Cabaret Voltaire, and NON (aka Boyd Rice). These cassette culture releases often featured zany tape editing, stark percussion and repetitive loops distorted to the point where they may degrade into harsh noise. In the 1970s and 1980s, industrial noise groups like Current 93, Hafler Trio, Throbbing Gristle, Coil, Laibach, Steven Stapleton, Thee Temple ov Psychick Youth, Smegma, Nurse with Wound and Einstürzende Neubauten performed industrial noise music mixing loud metal percussion, guitars and unconventional "instruments" (such as jackhammers and bones) in elaborate stage performances. These industrial artists experimented with varying degrees of noise production techniques.

Other postmodern art movements influential to postindustrial noise art are Conceptual Art and the Neo-Dada use of techniques such as assemblage, montage, bricolage, and appropriation.[102] Bands like Étant Donnés, Le Syndicat, Test Dept, Clock DVA, Factrix, Autopsia, Nocturnal Emissions, Whitehouse, Severed Heads, Sutcliffe Jügend and SPK soon followed. For me, their noise stood in defiance of the limits of ordinary perception and representation. Thus it was about the opposition between the various pleasures of standard music and the transgressive/ecstatic moment. In a sense, it attempted to set up a stable form of ecstatic transgression where I could go back and forth at will. This is perhaps similar to the tongue-in-cheek idea behind the amusing *Excessive Machine* in the film *Barbarella*.[103]

The sudden post-industrial affordability of home cassette recording technology in the 1970s, combined with the simultaneous influence of punk rock, established the no wave aesthetic, and instigated what is commonly referred to as noise music today.

Lou Reed's double LP album, *Metal Machine Music* (1975) is an early, well-known example of commercial studio noise music that the music critic Lester Bangs has called the "greatest album ever made in the history of the human eardrum". It has also been cited as one of the "worst albums of all time". Reed was well aware of the electronic drone music of La Monte Young. His Theater of Eternal Music was a seminal minimal music noise group in the mid-1960s with Velvet Underground cohort John Cale, Marian Zazeela, Henry Flynt, Angus Maclise, Tony Conrad, and others. The Theater of Eternal Music's discordant sustained notes and loud amplification had influenced John Cale's subsequent contribution to the Velvet Underground in his use of both discordance and feedback. John Cale and Tony Conrad have released noise music recordings they made during the mid-sixties, such as Cale's *Inside the Dream Syndicate* series (The Dream Syndicate being the alternative name given by Cale and Conrad to their collective work with La Monte Young).

The aptly named noise rock fuses rock to noise, usually with recognizable rock instrumentation, but with greater use of distortion and electronic effects, varying degrees of atonality, improvisation, and white noise. One notable band of this genre is Sonic Youth who took inspiration from the no wave noise composers Glenn Branca and Rhys Chatham (himself a student of La Monte Young). Marc Masters, in his book on the no wave, points out that aggressively innovative early dark noise groups like Mars and DNA drew on punk rock, avant-garde minimalism and performance art. Important in this noise trajectory are the nine nights of noise music called *Noise Fest* that was organized by Thurston Moore of Sonic Youth in the NYC art space White Columns in June 1981, followed by the Speed Trials noise rock series organized by Live Skull members in May, 1983. Also notable in this vein is *Unfinished Music No.1: Two Virgins*, an avant-garde recording by John Lennon and Yoko Ono from 1968 consisting of repeating tape loops as John Lennon plays on different rock instruments such as piano, organ and drums along with sound effects (including reverb, delay and distortion), changes tapes and plays other recordings, and converses with Yoko Ono, who vocalizes ad-lib in response to the sounds. They followed this recording with another noise recording in 1969 entitled *Unfinished Music No.2: Life with the Lions*.

Since the late 1980s in Japan, there has been a prolific output of "harsh" noise music (often referred to as *japanoise*) by the noise figurehead Merzbow (pseudonym for the Japanese noise artist Masami Akita who himself was inspired by the Dada artist Kurt Schwitters's Merz art project of psychological collage). Other Japanese noise artists include Boredoms, C.C.C.C., Incapacitants, KK Null, Yamazaki Maso's Masonna, Solmania, K2, The Gerogerigegege, and Hanatarash.[104]

Following in the wake of industrial noise music, noise rock, no wave and harsh noise, there has been a flood of noise musicians whose ambient, microsound or glitch-based work is often subtler to the ear.[105] Kim Cascone refers to this development as a postdigital movement and describes it as an *aesthetic of failure*.¹⁰⁶ Post-industrial noise artists from the 1980s, 1990s and 2000s include Alva Noto, Nicolas Collins, Boyd Rice, The Psychic Workshop, Signal, Stephen Vitiello, If, Bwana, PBK Phillip B. Klingler, Aube, Crawling With Tarts, Andrew Deutsch, Randy Grief, Robin Rimbaud, Minoy, Kim Cascone, Master/slave Relationship, Oval, Boards of Canada, Maybe Mental, Kenji Siratori, Thanasis Kaproulias (Novi-Sad), Fennesz, Matthew Underwood, Yasunao Tone, Noise Maker's Fifes, Pole, Arcane Device and Francisco López, among many others. Richie Hawtin, Jan Jelinek, Ricardo Villalobos, Decomposed Subsonic, Trentemøller and other Minimal techno and Microhouse DJs have been using noise elements such as buzz, hum and clicks as sonic flavor since the early 1990s.¹⁰⁷

Their noisy view of post-industrial society takes into account the rich ensemble of possible relations (the diversity, the unexpected links, the ruptures, the amalgamations, the connected heterogeneity) that Deleuze and Guattari showed us. Indeed for many artists, myself included, Deleuze and Guattari's vision of post-industrial life re-opened a way for the production of subjectivity in art¹⁰⁸ by affirming the befittingness of *multiplicity* coupled with the necessary right to *dissension* typical of art noise.

Tabulating Datamatics [ver. 2.0]

On October 29th, 2007 I attended a concert by the Japanese composer Ryoji Ikeda at the Centre Pompidou in Paris called *Datamatics [ver 2.0]*. At the time, *Datamatics [ver 2.0]* was the latest electronic audio/visual creation of Ikeda where he mines data mania for both the material and the theme of his work. The intention is a meditation on the wild relationship between the sound of data and the data of sound today. The effect, however, is a furious formalism that effectively entices, but flattens and thins out the longer it goes on. That said, the macabre grandeur of *Datamatics [ver 2.0]*, with its repetitive super-coded/anti-coded rigor, is stunningly beautiful on début. A furious rhythm of inscrutable data discord is established from the beginning, necessarily entailing a process of attraction/repulsion that intimidated me while spawning some sublime ideas.

Ikeda makes speed manifest here, including the various speeds and slownesses that extend the retinal limit in a way that would be previously regarded as outside of phenomenological thought. The complex installation work of Tatsuo Miyajima came to mind at points. Specifically Miyajima's 1996 installation at La Fondation Cartier pour l'art contemporain, where he made two large installations which dealt with the abstract constitution of time in the digital age. Both installations consisted of abundant LED signal-lights that flashed a countless bevy of over-excited digital numbers in what appeared to be a random order. One installation, *Time Go Round*, had twenty green and red digital modules spinning in various circular orbits against an imposing dark wall. One discerned there a mystifying data constellation in transit, reminiscent of passages from *Mona Lisa Overdrive*.[109]

Time Go Round was an attempt to delineate the crisis of time in relationship to the dispersed ontological self in the information age (where digital time as the only time has become non-problematic in computational work environments). Miyajima's artistic sense of time in crisis served to encourage me to value the freedom of my own interior sense of time.

By contrast, Ikeda's evocation of data time is riding high on speed, and tempo here took on the implication of a dark temporal pop-cultural product pit into which my accurate perceptions were poured—even as I resisted fragmentation and remained fixed in the logocentric seat of Renaissance three-point-perspective. This principle of hyper speed coupled to visual overload makes inoperable the usefulness of the term 'minimal' in association with Ikeda, as *Datamatics [ver 2.0]* animates a crumbling of the normal monuments to human difference we construct daily. Ikeda's mixture of technical precision with perceptual overload presented a significant challenge to experiencing interior time. Perhaps it would have been possible had I been able to divorce the musical experience from the visual torrent.

Ikeda's rapid techno music is created from slight electronic hums and pops that build into gargantuan sonic textures, sometimes reaching the noise intensity of Merzbow. Given his cornucopiastic range, Ikeda, quite scrupulously, defies melodious categorization. This range allows for virtuoso moments that provide the opportunity of exploring the intricacy of his hard-edged myriad-colored dexterity as he plays back-and-forth with elaborate but lucid musical aggregates that facilitated mild waves of aural imbrication. In piercing clouds of cacophony I heard traces of Xenakis, La Monte Young, Boulez, and Aphex Twin.

But Ikeda's primary tool of coherence is what in acoustics is called 'duration', the steady-state of a sound at its maximum intensity. My supposition here is that Ryoji Ikeda takes this musical phenomena of duration and extends it into a general spatial intelligence based on petite bursts of sound. The attraction to such an adjoining structure is strong, but it wanes quickly. I suppose it must be like doing business with a rather spectacular whore.

Conceptually, Ikeda's music reminds me that our once basic Euclidean conception of space has been expanded to include the formation of many-dimensional space. In Ikeda's music, the Euclidean concept of space is modified via excess by enlarging the number of vectors that may be constructed within it from three to some much larger number (designated as n). Such n space implies the existence of a higher-dimensional geometry that mimics Euclidean geometry. Inevitably this approach shaped me as the viewer/listener into an inert subject. The audio is both clean, noisy and hardheaded in such a way that the individual's personal extension into the virtual tends to be blunted.

There is also, however, another proposed spatial reality relevant to Ikeda, most notably the topological space model of fuzzy space where there exists only a concept of nearness. In this respect he reminds us that hearing and seeing is not an activity divorced from consciousness.

But really, any account of Ikeda's sonic dexterity as related to consciousness is inadequate to our actual experience of it. Yes, his music is conceptual in that his sound deprives us of our habitual perceptive boundaries by surpassing them. Through the excessive, Ikeda makes us remember that throughout time there have been consensual realities that have proven to be nothing but vast daydreams. But Ikeda's music spectacularly fails to be in opposition to what Donald Lowe in his *History of Bourgeois Perception* identifies as the "bourgeois perceptual field", a mode that he characterizes as fundamentally linear, non-reflexive and overtly objective.[110]

So, to conclude, Ikeda's initially mesmerizing presentation was an experience always about to come. I say 'about to come' as *Datamatics [ver 2.0]* contains much manic machinic stuttering (full of Nietzschean multiple affirmations and shattered teleological art-historical/art-hysterical continuums) that never resolve. The stuttering I am addressing here rests, of course, in the spectral repetitions of his mental-machinic procedures:

we see and hear a digital/mechanical shifting again and again and again and again and again and again, but with slight variations full of dazzling élan. So it is a vigorous abstract stuttering I sensed in the work that took me down into a deadening sensation of unfathomable data: the data of anytime-anywhere. Given that implication of mythic indifference, Ikeda's a/v stuttering could be properly aligned with the Dada artistic legacy of Hugo Ball and Tristan Tzara. It is a form of digital-dada, post-conceptual art-music in its absurd machinic indifference.

How does *Datamatics [ver 2.0]* achieve this meticulous indifferent stuttering? Ikeda uses the precision of digital technology to fracture data into tangled networks of beeps and lines—initially delightful and exquisite nihilistic manipulations that tease our mind with their multiple syntactical/semantic gestures of sadism, strenuously massacring the social source material along the way. But like Op Art[111] (which it resembles) on crack, this stuttering stuttering stuttering turns tedious and cold, shutting down feeling, reflection and contemplation and hence imagination in my mind. In that sense, Ikeda only created pictographic and aural excavational moments that cannot be sustained, but are instead mental acts worthy of short but frequent revitalizations: again and again and again and again—a visual/audio whiplash that slashes into the burnt annals of symbolist romanticism. To follow Ikeda there is to evaporate into the puzzling archives of some geek heretical doctrine and pop out again into a dead excess vis-à-vis ideology writ large as system. In that sense, he pictures/sounds as an obscene thrashing of, and ongoing onslaught against, innocence.

Alors? So are these subsequent revelations an abiding labyrinthian form of abject nothingness? Yes, *Datamatics [ver 2.0]* is a blustering, bursting, blatant, banality, but even so I saw/heard in *Datamatics [ver 2.0]* the melancholy monstrous traces and dissimilative Dionysian mannerisms of Novalis, Chateaubriand, Nerval, Baudelaire, Rimbaud, Aragon, Bataille, Lautréamont and Roussel. That is hardly farcical nothing (albeit by way of negation folded upon negation/instrumentalization upon instrumentalization). But what is missing in *Datamatics [ver 2.0]* for me is vague imagery and sound that does not depend on induction or deduction, and exists prior to these forms of controlling cognitions. In that sense, *Datamatics [ver 2.0]* cries out for access to the libraries of other people's subconscious experiences.

So Ryoji Ikeda's *Datamatics [ver 2.0]* is a stuttering in a hygienic but deranged tongue within the vernacular of shattered techno signs and computer music clichés. In that way *Datamatics [ver 2.0]* is anti-automatismic. We are forced to think creatively and distinctively if we hope to un-pack and self-interpret its quintessentially dancing chaotic vision par excellence. And when we do: we finally do come—enigmatic-lithe jouissance. But the jolt has been sadly self-inflicted, lacking, as it does, the tragic/emphatic psychic dimensions of artificial life (I saw or felt no field of intensities invoking the inchoate and the savage) and the open multiple model of atmospheric free associations. Thus the event went a bit lacking, for me, in what Deleuze suggested to us (via Proust): something real without being actual, ideal without being abstract.

Tabulating Cecil Taylor

Ikeda's concert brought back to mind another concert I attended by Cecil Taylor at Alice Tully Hall in New York on the 28th of February in 2002. Given his gigantic musical range, Cecil Taylor quite scrupulously defies melodious categorization. What we can say is that, with a piano, Taylor creates gargantuan sonic envelopes to float in. This I eerily experienced again at his euphonious performance as he exquisitely explored the intricacy of his myriad-colored dexterity by playing back-and-forth with elaborate/lucid musical paradigms and triumphal aggregates like the virtuoso he is. In clouds of muscularly produced tinkling "cacophony" were heard exquisite dashes of Fats Waller, Xenakis, Bud Powell, La Monte Young, Brahms, John Coltrane, Aphex Twin (drukqs), Sun Ra, John Cage and, of course, Theolonius Monk. His music, then, is a vigorous paradox where customary opposites coexist, coalesce, and connect.

Cecil began recording in 1955, working with Steve Lacy, Buell Neidlinger, and Dennis Charles. (Neidlinger, an experienced symphonic bassist, once said that no musician he'd ever met, including Stravinsky and Boulez, had musical abilities that exceeded Taylor's, and that he is potentially the most important musician in the western world.) The group had an extended, six-week engagement at the Five Spot Cafe in New York—literally introducing the concept of modern jazz to a club that shortly thereafter became one of its legendary venues—and made an appearance at the Newport Jazz Festival, which Verve recorded in 1957.

In 1958, he recorded with Earl Griffith, Charles and Neidlinger for Contemporary (*Looking Ahead*), and with John Coltrane for United Artists (*Coltrane Time*). Taylor also made his own date for that label in 1959, and more Candid sessions through to 1961 (Mosaic collected the complete Candid sessions on a comprehensive four-disc boxed set in 1992).

Taylor's primary tool of coherence is what, in acoustics, is called *envelope*. Envelope, in music, involves the onset, growth, and decay of a sound. Growth consists of the rate of increase of a sound to steady-state intensity. Duration refers to the steady-state of a sound at its maximum intensity, and decay is the rate at which it fades to silence. Envelope is an important element of timbre, the distinctive quality, or tone color, of a sound. My supposition here is that Cecil Taylor takes this musical phenomena of envelope and extends it into a more general peripheral spatial intelligence best called holonogic.[112] Yes, I think it is sensible to make use of the holonogic schematic model of Arthur Koestler (1905-1983) (established in his books *Beyond Reductionism* and *The Ghost in the Machine*)[113] when trying to appreciate the music of Cecil Taylor in that no set or frame of perceptions may be experienced in isolation or as a single part of a finite perceptual collection within the holonogic model.

Taylor's performance fantastically fits the holonogic paragon starting with its ritualistic beginning (which suggested deep African ceremonial consciousness in dialogue with Taylor's roots as a tap dance) to its empyrean conclusion.

The concert began with four intense piano solos, with Cecil producing some muffled and deeply eccentric vocalizations. These solos all imploded and exploded with those detailed musical references cited above. I had to close my eyes to even attempt at hearing all the musical ideas simultaneously present. The acoustics are phenomenal at Alice Tully Hall, though, and they facilitated mild waves of aural imbrication. This made for a rather complex musical reckoning.

Taylor's music reminds me that our once basic Euclidean conception of space has been expanded to include the formation of many-dimensional space. In Taylor's music, the Euclidean concept of space is modified by enlarging the number of vectors that may be constructed within it from three to some much larger number. There is also, however, another proposed spatial reality relevant to Taylor's music called curved-space—or curved space/time. Curved-space is approximately Euclidean over very small regions, but over large regions all geometrical properties break down. Curvature is combined with Euclidean geometry with the increase of dimensions plotted. There are also a number of other generalized spaces that drop the Euclidean geometry completely, most notably the topological space model and fuzzy space where there is only a concept of nearness.

Part II of the concert consisted of Taylor's piano playing with (sometimes within) the dazzling percussion of Jackson Krall. This accomplished performance—which included him sassily playing the mise-en-scène with his sticks—would be appropriate to anybody analyzing noise music and the holonogic principle. The colossal, deep base atmospheric spectrum was handled by Dominic Duval. Playing concurrently, the three sensitively roared. At moments I could hear the majestic universe bellow—and then whimper. Everybody who has ever seen Taylor play live—with or without other musicians—knows this. He does this to the entire room by unframing our mind and ears and expanding the listener's sensitivity both to noise and to the most delicate tiny musical moments. The music then remains beautiful to recall. In this respect he reminds us that hearing is not an activity divorced from consciousness.

But like with Ikeda, any account of Taylor's sonic dexterity as related to consciousness is again inadequate to our actual experience of it. But the holonogic model of cognitive-aural processing is useful for a one-of-many possible accounts of its reverberations. Yes, his music is particularly holonogic as his sound deprives us of our habitual perceptive boundaries by surpassing them. Through excessive deprivation, Taylor makes us remember that throughout time there have been consensual realities that have proven to be nothing but vast daydreams, such as the conviction that the earth is at the center of the universe. Yes, the holonogic model befits Taylor's adroitness because according to Koestler's holon concept, instead of cutting up immersive perceptual wholes into discrete focal parts, noise immersion should be scrutinized and understood using synthetic sub-whole sets found within ambient space.[114] And Taylor's music deserves this level of attendant complex scrutiny.

Such an approach to Taylor's music is consistent with, and indeed epitomizes, the ideals of hermeneutics, insofar as in hermeneutics the central notion is that we cannot grasp the meaning of a portion of a work until we understand the whole, even though one cannot understand the whole until one understands the parts that make it up. However, hermeneutics is not merely a paradox, since hermeneutics indicates that any act of interpretation occurs through time, with adjustments and modifications being made to one's comprehension of both the parts and the whole in a circular manner, until some type of resolution is attained.

Such an extensively engrossed holonogic/hermeneutic approach towards the music of Cecil Taylor is noise in opposition to the bourgeois perceptual field. Insofar as our adult creativity derives primarily from our conspicuous potential for abstraction (which characterizes our genus) and in our craving and manipulation of abstractions, what is at stake for Cecil Taylor here is our acceptance of our entire atmospheric sensation as our genuine field of conscious creative interest—an abstract field that calls on our tremendous expansive qualities for which the descriptions of the scientist and the doctor have not done suitable justice.

Tabulating Power Electronics

My increased involvement in this topic of noise music was launched in downtown Manhattan during the heyday of no wave music. My enjoyment of the music of Glenn Branca, DNA, Rhys Chatham (with whom I played music, but more importantly collaborated with on a no wave noise opera called *XS*)[115] and others, coupled with my growing knowledge of Fluxus and Minimalist art music as archivist to La Monte Young, led me to curate two noise issues of *Tellus Audio Cassette Magazine*:[116] *Power Electronics* (1986)[117] —which led me to meeting Merzbow, and later, *Media Myth* (1988).[118]

The basic premise behind *Power Electronics* and *Media Myth* was the exploration of the introspective world of the ear under the influence of the era's high-frequency electronic environment. Since it was difficult making sense of the 1980s' swirling media society, the general proposition behind *Power Electronics* and *Media Myth* was to look for a paradoxical summation of this uncertainty by looking for artists who took advantage of the time's superficial saturation—a saturation so dense that it failed to communicate anything particular at all upon which we could concur, except perhaps its overall incomprehensible sense of ripe delirium—as the Reaganomic reproduction system pulsed with higher

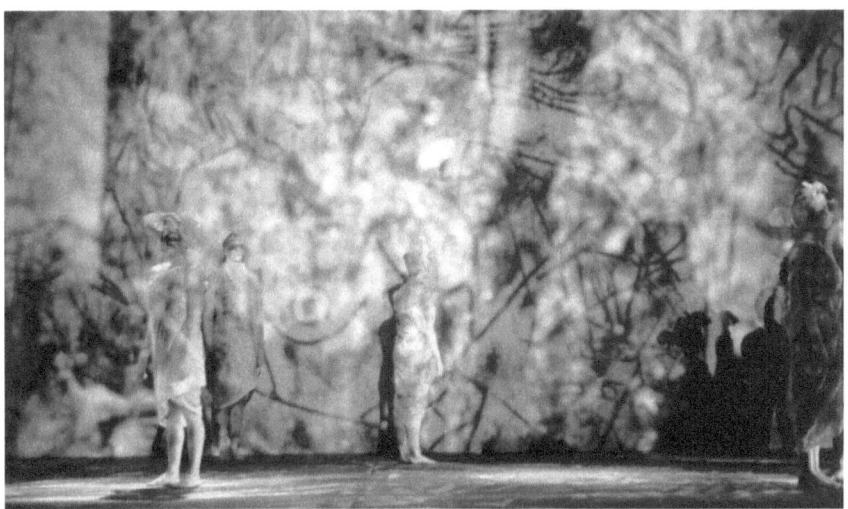

Figure 3 – *XS the Opera*: Shakespeare Theater Boston 1986

and higher, faster and faster, flows of senseless info-data to the point of near hysteria.

Perhaps the result of this ripe information abundance was that the greater the amount of Reaganomic information that flowed, the greater the incredulity which it produced, at least for thinking, questioning artists. So, the tremendous load of data produced and reproduced all around us ultimately seemed to make less, not more, conventional sense. Indeed, this feeling became the premise of both *Power Electronics* and *Media Myth*.

This supposition about noise and noise music, it seems to me, plays into the history of abstract visual art which teaches us that art may refuse to recognize all thought as existing in the form of representation and that, by scanning the spread of representation, sound art may formulate an understanding of the laws that provide representation with its organizational basis. As a result, in my view, as mentioned above, it was electronic-based sound art's onus to see what unconventional, paradoxical, summational sense—in terms of the subjective world of the imagination—art might make based on an appropriately decadent reading of our time's paradoxically material-based (yet electronically activated) media environment.

Such a basically abstract, artistic, paradoxical/summational fancy began with the presumption that an information-loaded nuclear weapon had already exploded, showering us with bits of radioactive-like informational sound bytes, thus drastically changing the way in which we perceive and act, even in our subconscious dream worlds. It is this internal, subconscious, paradoxical drama of the Reaganomic, 80s—this subconscious contradictory tension—that I found potentially interesting in conceiving *Power Electronics* and *Media Myth* and as a subject specifically suitable for electronic-based sound art.

Electronic-based sound art, by virtue of its distinctive electron constitution of fluidity, floats in an extensive stratosphere of virtuality. Hence, the particular constitution of *Power Electronics* and *Media Myth* is best seen, perhaps today, as an osmotic membrane: a blotter of the 1980's instantaneous ubiquity/proliferation. Consequently, *Power Electronics* and *Media Myth* reflected (and worked with) social power. *Power Electronics* and *Media Myth*, when viewed as shaped by the de-centered electronic overload of the 80s, is understood today as a flustered code of digital signifiers, a confused collective representation that bewilderingly mutated the ideology of its own reproduction.

So, the question for these projects was, how could artists symbolically turn these de-centered power codes into artistic abstractions of social merit? Perhaps it was possible because I knew, somewhere, that these symbolic codes—which, after all, helter-skelter, make us up—are positively phantasmagorical.

This is still, of course, true today. Based on the premises of *Power Electronics* and *Media Myth*, perhaps a socially relevant digitally-based audio noise creativity can be found also in today's electronic-based art's ecstatic potentiality—in that electrons partake (and make up) the all-encompassing phantasmagorical/technological sign-field in which we live and which defines us (at least in part). Since prevailing representation is made up of conventional, rigid social signs (and sound art typically of unconventional irresponsible signs—the mode that represents the real arbitrary nature of all signs as it subverts the socially controlled system of meaning), electronic-based sound art may offer us the opportunity for the creation of relevant and applicable anti-social phantasmagorical signs (hence, abstract ecstatic anti-signs) which may continue to mentally move and multiply us without stop. Yet this fancied aesthetic non-knowledge is certainly the most erudite, the most aware, the most conscious area of our current identity, as it is also the phantasmal depths from which all digital representation emerges in its precarious, but glittering, existence. Indeed it was this quivering phantasmal cohesion that maintains the sovereign and secret sway over each and every audio sample—this phantasmal vibrating—which I found interesting in conceiving both *Power Electronics* and *Media Myth*; a sway beyond reductive minimalist abstraction into an excessive hybrid abstraction—an audio art, which is in theory opposed to the tabular mental space laid out by classical thought.[119]

Surely such a hybrid electronica/phantasmal impetus can help release pent up ecstatic energies in that the more overwhelming and restrictive the social mechanism, the more exaggerated are the resulting effects—and hence excel the assumed determinism of the technological-based phenomenon inherent (supposedly) in our post-industrial information society. Therefore, in this way, I hoped *Power Electronics* and *Media Myth* served as an ecstatic impulse/phenomena that proliferates in proportion to the technicization of society—as such a nervous electronica-ecstasy may occur as a result of technological society's obsession with the phantasmal character of electronic speed-proliferation.

My contention is that, as human psychic energies are stifled and/or bypassed by certain controlling aspects of mass technology, such a nervous ecstatic phenomena will increasingly break out in forms of noise art. Similarly, simulation technology—when used in the creation of electronica-based art—promotes an indispensable alienation from the socially constructed self, necessary for the outburst of such nervously ecstatic experiences/acts. Inversely, electronic technology enables the contemporary artist to express nervous ecstatic reactions in ways never before possible. Thus, this nervous ecstatic counteraction provides a phantasmal defiance through transport aimed against the controlling world's blandness and self-destructiveness. This aesthetic noise philosophy provides a fundamental antithesis to the authoritarian, mechanical, simulated rigidities of the controlling technical world.

May I just say that the nervous phantasmal play of noises found in *Power Electronics* and *Media Myth* (using strategies of jamming, non-communication, miscommunication, interference and disruption) has urgent political/social ramifications in our media-saturated society today. The artists' well-founded but ambiguous phantasmal model for noise art indicates the capacity for electronic media to jolt consciousness, as it provides the explication of the nervous phantasmal links that abet communications. Hence, excessive and noisy audio abstractions, such as those found here, can be (in a sense) the representation of all representation when they attempt to represent the unlimited field of representation via noise, a noise which non-utilitarian phantasmal ideology attempts to scrutinize in accordance with a non-discursive method. An unlimited field of representation that now appears as an abstract din of nerve-noise where excessive abstraction helps us to step outside of representations to posit us outside of the mechanics of uniform and anthropomorphic dogmatism.

With this art of noise as an unlimited field of representation in mind, we will now open our eyes.

CHAPTER 2

Noise Vision

*...'consciousness' in the function of self-reflexivity
should be operating within the elements of
the work (proposition) of art itself.*
— Joseph Kosuth, *Within the Context: Modernism and Critical Practice*

*Lascaux is the passage from the work world to the play world,
which is the passage from the Homo Faber to the Homo Sapien.*
— Georges Bataille, *Lascaux: La Naissance de l'Art*

The transition from audio noise to visual noise based on ideas of an unlimited field of representation requires, I believe, the judicious use of the process of Deleuzian/Guattarian nomadic thinking. Accordingly, Deleuzian/Guattarian noise descriptions would be composed of variously formed segments, stratas, and lines of flight that involve territorializing as well as deterritorializing spacio/psychic activities.[120] Even so, I acknowledge in advance that all methods, explanations, and theories (including the nomadic) inevitably distance consciousness from its first sense of full and total participation. This acknowledgement will remain a particularly important point of consideration here, as ideas of spacio/psychic critical distance and non-distanced (non-spatial) disembodied fusion rub up against each other and influence the psychic space required for reflection on the thorny concept of aesthetic immersion into noise (which entails a lack of distance) as the atmospheric gulf between the immersant and the immersive aesthetic environment is ideally dissolved.

In this light, it might be possible to define noise art as *conditions and orders of conscious awareness in which perception-cognition (i.e., awareness linked to the process of forming intelligence) is found to consist of* more *than everyday (non-conceptual) vision or hearing typically reveals, by merging it with some manifestation suggestive of a magnificent more*. This condition may be thought of as a bypassing of habitual processes of art through an assiduously expanded macro-intelligence based on conditions of excess that provide us with an unfilled sense of internal union with unrealizable breadth through noise.

When I use the terminology *expanded* here I am referring to the rich meaning given to it by Gene Youngblood in his book *Expanded Cinema*, as that which transgresses and exceeds the customary boundaries of our optic encounters. When Youngblood discusses what he calls *expanded cinema*, he refers it to an "expanded consciousness".[121]

By *noise consciousness* I mean, then, our miscellaneous neurological/ontological sense of the gradient unity of sentient self in internal discord[122] with its surrounding milieu, that mental property of atmospheric self-attentive awareness, cognisance and feeling that allows us to experience a sense of nexus with our ostensibly unified surroundings, albeit laced with vicissitudes.[123] I have observed (in myself) that noise in art tends towards unconstrainment while being based on a routine sense of shifting-self within the ambient scene which is experienced when self-attentive.

From a philosophical perspective, *synthesis* is the procedure by which, once thought, separate elements of a system are assembled into a union of an undivided whole, so that the consequent unity is something more than the mere sum of its unmitigated parts. Synthesis proceeds from the stand-alone, separate elements discerned by analysis, but it supersedes analysis by raising the particulars to the point of being conscious of their larger comprehensive framework.

According to Theodor Adorno in his *Aesthetic Theory*, art and aesthetics must not try to erase fractures through integration but rather "preserve in the aesthetic whole the traces of those elements which may have resisted integration".[124] Noise as art does just that. However, an understanding of this noisy, self-attentive shifting-self through listening requires a surpassing of the limiting tropes of logical positivist empiricism (typical in indexical thought), as noise consciousness starts in the non-delineating darkness of closed (but debonair) eyes. That is where the beginnings of idiosyncratic human imagination seems to dwell—in the dark. But a consideration of this self-attentive, immersed, shifting-self is also post-logical positivist in that it accepts various theories of consciousness that discuss consciousness as being emergent rather than representational. Sigmund Freud (who we must remember was a theorist who rooted his theories in anecdotal evidence and whose writing was literary) identified an artist as one who offers insights into such an emergent consciousness as it emerged from within the unconscious realm. Moreover, Martin Heidegger (1889-1976) maintained that *being* is the most unconscious of concepts because we are thoroughly immersed in it.

The terminology *consciousness* means verbatim *with knowingness* and stems from the Latin verb *scire* (which means *to know*), as does the word *science*. But this is not all there is to it as applied to art. For consciousness in art seems to be ultimately like a web woven in the mind/body, of various silken-strands spun forth from interlacing states of unconscious desire which semi-automatically control the paradigmatic creation and reception of art. This definition coincides with R. G. Collingwood's definition of consciousness, in paradigmatic art terms, as that which is a "kind of thought which stands closest to sensation or mere feeling" as "transformed into imagination".[125] Paradigmatic consciousness has emerged in the 20th century largely due to the philosophical work of the American philosopher Thomas Kuhn who argued that scientific "progress" does not simply occur in stages based on neutral observations but that all observation is theory-laden. For Kuhn, the history of science (and I would argue art as well) is characterized by revolutions in outlook.[126] Indeed, unconscious desires shape the paradigms that contour intentional expressions in art through the subtle powers of sublimation when the sexual desires of the libido are turned into cultural ones via the mediation of the artist's ego. The question of how Freudian unconscious desires are manifest in conscious cultural noise production and interpretation will be one of our minor themes throughout. This is a non-problematic working assumption in that even those who maintain that art is fundamentally a materialistic, social, and conscious product[127] acknowledge that the role and function of art is located in its power to change consciousness.[128]

Nerve Noise Visualization

In the realm of the affective imponderable, the image provided by my nerves takes the form of the highest intellectuality, which I refuse to strip of its quality of intellectuality.
— Antonin Artaud, *Manifesto In Clear Language*

One way to apprehend an ambient field's *felt scopic atmosphere* is to think of it in terms of a study of *cognitive-visual acoustics*. This is equitable in that sight itself is nothing other than a continuous pattern of perpetually changing light-data recorded on the retina which we humans process through the aggregated internal acts of discerning. To understand noise vision as being non-inflected with subtle properties akin to the acoustic properties of echo, range, pitch, timbre, and tone is to discern all visual moments as being indiscriminately equal, and as flat. Cognitive-perceiving is continuously allocated by tones of recognition, ranges of totality, and distributed visual echoes as humans produce a full interpretation of the plethoric information which hits their retinas in order to assign it cultural meaning. More precisely, such an acoustic-like cognitive-visuality would involve the equivalent of what in acoustics is called *envelope* (previously explained) as visual attention has characteristics of attack, growth, duration, and decay in terms of peripheral spatial intelligence (when self-attended to). Such attention calls for the viewer's active and self-conscious engagement with art.

By studying such an *envelope noise vision* in terms of immersion, in a sense this book participates in the investigations of visuality into what Martin Jay has called the "ocular character of all Western culture" and the "Cartesian perspectivalism that dominates the modern era"[129] — a Cartesian perspectivalism which, according to Hal Foster, separates subject from object, "rendering the first transcendental and the second inert".[130] Such investigations include Guy Debord's (1931-1994) critique of the *Society of the Spectacle*,[131] Jacqueline Rose's investigation into the sexuality of the objectifying, male, patriarchal gaze,[132] and Michel Foucault's (1926-1984) analysis of the panopticon paradigm.[133]

When talking about noise vision, it must be remembered that, in philosophy, synthetic statements are those statements judged to be true or false in relationship to the world (but which are not necessary ones), as opposed to analytical truths, which are necessary, and hence cannot be otherwise. In philosophy, it is important to make this distinction between synthetic and analytical statements. Only when we acknowledge that this investigation of *noise vision* partakes in synthetic activity might we enter the concept into consideration, and only if we understand noise vision to be a synthetic *psychological thought-vision* without any one particular vector but rather a plethora of them united into one *void of the suppositional central vanishing-point which the horizon-line had previously established*.

The synthetic notion being pursued here, then, is of an atmospheric noise vision constituted by what goes on in and behind the head as much as by what is in front of it.

As Jane Ellen Harrison (1850-1928) tells us in *Ancient Art and Ritual*, art is not mimesis, but rather mimesis comes from art's emotional expressions.[134] We must additionally recognize that cognitive noise vision takes place not only over time but within the emotional brain and that much of noise consciousness is supra-sensible. Particularly germane to our inquiry into noise vision is the fact that most aesthetic theories argue that art is not a matter of simple embellishment considering its diverse appeals to the various cognitive faculties of the eye/mind complex. The critical capacity of art is that it advances conjoined expectations along with cultivated appraisals through discriminating semi-withdrawal. So considered, assumptions concerning noise vision in sacred zones (and their distinguished semi-removed status) in regard to immersively spawned states of aesthetic consciousness will be addressed historically in this section, as it is common for ideal sacred zones to supply acute information on the human race's apparently insatiable desire for transcendence through immersive aesthetics.

In this respect, I go along with Georges Bataille as he argued that the sacred springs from the same sources as those things we conventionally find repugnant, such as noise, ritual sacrifice and bodily mutilation, and that within sacred zones sublime transmissions are meant to transpire, thus provoking attachments between the unconscious mind and its conscious active comportment. The marvellous abstract character of such supposed sublime transmissions in terms of noise vision will be explored in this section. But, to begin to do so, we must keep in mind that all reputed sacred propositions occur within configuring theories of culture. All that we apprehend as sacredly significant resides in cultural symbol, which is the gist of art. It is exclusively by our encounters with theories of culture that we style omitted or grasp upheld sacred abstractions.

With this in mind, we shall now turn our attention to what I perceive as the genesis of immersive noise vision: the adorned prehistoric cave. We shall approach the resplendent prehistoric cave by keeping in mind that, according to Marshall McLuhan (1911-1980), form determines the action of mediation which determines meaning.

A prehistoric *painted cave* is all that, enhanced moreover through the emotional defamiliarizational powers of art. Over 200 late-Stone Age caves bearing wall paintings, engravings, bas-relief decorations and sculptures have been found in south-western Europe alone. Life, in the form of tiny blue algae, emerged on Earth 2 billion years ago, or what is called *BP* (Before Present). The first people who made tools, the basis of technology, were the Homo Habilis, a people who lived in Africa 2 million years ago. People have inhabited the Périgord region of France for about 200,000 years and indeed the cave at Lascaux was discovered by Cro-Magnon people.[135]

Gradually during the Gravettian Period (approximately 20,000 to 25,000 years ago), people began to embellish the walls and ceilings of a few small shallow caves.[136] Subsequently, prehistoric painted caves became the sites of these humans' first topographical imagings, images that celebrated mortal terror and love of the animal and its world, as well as the passionate and jubilant triumph over that terror/love through the organized hunt and the strategic, co-ordinated, co-operative group adhesion which the hunt necessitated.

However, it is important to remember at all times that the animals depicted in the caves were not generally those animals that were hunted and eaten. The Magdalenian people hunted and ate primarily reindeer and a reindeer is only represented once in the cave of Lascaux out of over 2,100 legible images—in the Apse.[137] The significance of this will be pondered and discussed shortly. But, at the outset, we can surmise that the animals represented here were depicted in order to serve as spiritual intermediaries or as ideal aspirations. In the terms of noise vision, the Magdalenians' depicted events can be interpreted as disassociated (in their lack of depicting context) and conflicting (in their superimpositionality) while being immersed underneath at the implicate frequency level, as these scenes depict all things and events as ultimately intangible and connected into one total singularity. It is for this reason that the prehistoric painted cave must be addressed as a place of active immersive cognizance and not as a mere receptacle of discrete utilitarian (magical) images in service of the hunt in any simplistic one-to-one fashion, though some sort of indirect connection to their hunting culture is hard to repudiate, especially after the discovery in Lascaux of a large number of broken spearheads, all of which were engraved, often with a double interlocked herring bone pattern and a star with six rays.[138]

Most prehistorians agree that visual communications came into being somewhere around 40,000 years ago, about the time when Cro-Magnons reached Ice Age Europe and began decorating their tools and bodies with symbols. Living in small groups, they constructed tents from skins and huts from branches; however, (evidently) they possessed an incredible yearning for deep immersive experiences within the dark places of caves. Thus, in the caves they embellished, it is possible to see an immersive presentation in a collective space, a space which was not the property of any individual. This expansion from the decoration of the body to the cave is in itself an extraordinary act of immersive intelligence. The period between the invention of drawing, when animal forms and human genitals were engraved in rock 35,000 to 40,000 years ago by the Cro-Magnon on the banks of the Vézère, and the creation of Lascaux, is as long as the period of time separating us from the civilization of Lascaux. As much time elapsed between the first ornamental body and the cave paintings of Lascaux (about 17 millennia) as separates Lascaux from the first TV broadcasts. Nevertheless, Stacey Spiegel sees the Lascaux cave as being "the first *total art*",[139] and Howard Rheingold speaks of Lascaux as the first virtual reality.[140]

The physical and psychic risks involved in such a seemingly non-essential activity as painting inside a cave indicates that it was done, and indeed savoured, for some perhaps sacred antediluvian reason deemed essential enough to fashion an immersive space where human consciousness could plunge into extraordinary immersive experiences. The real threat implicit in the dangerous passage that must be made to enter a painted cave, with its usual remoteness from human habitation, suggests that these are sites of ritualistic loss and re-finding typical of intense love and tragedy. Thus the entrance into an immersive cave is always a movement towards self-interiority. To enter a cave is to move into it and, as such, initially involves a directedness away from the periphery and toward depth, toward noisy density, and away from dispersion.

Thus, far away from the light of the sun and stars, far from the daylight world of accustomed life, prehistoric people must have entered the depths of the immersive darkness of a cave to contemplate both the beginning and end of their life. Indeed the cave's lack of light is an insubstantial force whose intensity around the immersant must be carefully considered. The first occurrence we must contemplate in this regard is the dilation of the eye's pupil as entree to a dim cave is achieved. Noticeable is that in terms of vision and light and sex, the pupil's dilation indicates sexual attraction and facilitates it.[141]

Salient here is that the retina registers a field of 160 million points of light. The remarkable richness of natural light is due to the fact that it is a unification of focused and diffused light. Issues of light are issues of clarity and obscurity, issues which constantly vie with one another with an exacting power. The sun, which is roughly 57 million kilometers (about 93 million miles) from the earth, functions as the source of all light of course, but we must recollect that its effects are invariably qualified to a greater or lesser degree by the earth's atmospheric envelope through which the light must penetrate. The regular waxing and waning of light is often dramatically altered in its character and intensity by the apparent vicissitudes of changing atmospheric conditions. In order to realize how essential this combination of direct and diffused light is to our sense of well-being, one need only recall the deadening aftermath of a heavy overcast day when the whole world seems to be enshrouded in a pervasive melancholy.

The early-Upper Paleolithic period[142] saw significant innovation in stone tool technology and weapon systems by the early members of our species. Their invention of sharpened flint blades made the creation of almost all of their art possible, via carving and engraving. In painterly terms, the principal techniques of Cro-Magnon art involved brushes made of vegetable fiber or animal hair, tufts of fur, and the use of fingers, along with a blowing of pigment dissolved in saliva onto the wall.[143] The European predecessors to the Cro-Magnons were the strapping Neanderthals who successfully occupied Western Eurasia from about 200,000 BP up until they were superseded by the Cro-Magnons, sometime around 40,000 BP. Neanderthal culture, known as Mousterian, shows scant inklings of visual representation. However, there are traces of immersive symbolism in their burial sites as the corpses were surrounded by pebbles and bones with fragmentary patterns scratched onto them. Sometime after 40,000 years ago, at a time when the remaining Neanderthals shared the European landscape with the first Cro-Magnons, there was a relative explosion of ornament and graphic imagery among the earliest Cro-Magnons.

By the Upper Paleolithic period, Homo Sapiens had firmly established their existence based on hunting, fishing and the gathering of plants. In terms of art, the Cro-Magnons left behind dozens of sculpted ivory animals, moulded and fired clay statuettes, hundreds of engraved images on limestone blocks and cave walls, thousands of scrupulously decorated personal body ornaments consisting of ivory, shell, soapstone and animal teeth, along with the numerous and widely distributed female (so-called Venus) figurines. The earliest substantial body of surviving material relating to human sexual culture is the art of the Eurasian Upper Paleolithic, including its paintings of half-bestial males with erections, rock-cut vulvas, carved phallic batons, and the previously mentioned super-endowed nude Venus female figurines. These sculptural miniature statuettes of extraordinarily big-breasted human females are understood as contemplations on sex and fecundity[144] and a longing for oceanic unity and totality. The Venus figurines are entirely in the round and unconstrained from any physical site, thus hand-holdable and portable. Wonderful examples are the ivory Vénus de Lespuge from Lespuge, France (circa 27,000 BC) and the eyeless and bulbous stone Venus of Willendorf (circa 30,000 BC) which was found in Austria.

Evidently there was adequate time for spiritual-artistic acumen in the hunter-gatherer society, as case studies from various parts of the world show that sufficient food can be obtained with an average adult hunting, fishing and gathering (in common cause with others) in only three to five hours per day, less than people generally work now in our (so-called) advanced western civilization. The leisure time of many hunter-gatherers seems to have been abundant, affording adequate time for the fashioning of the immersive artistic/spiritual cave spaces which concern us here. Indeed, André Leroi-Gourhan in his book, *The Dawn of European Art: An Introduction to Paleolithic Cave Painting*, maintains that the generations of artists who executed Lascaux were very probably released from even this minimum burden of daily work by other members of the group.

Although Paleolithic cave art is often discovered deep inside caves quite remote from the cave entrance, it is a mistake to suppose that Upper Paleolithic human communities usually lived in such dark, and inherently hazardous, sites. Customarily, they lived in the open air, enjoying the sun and breeze, under skin tents or in the mouths of caves or beneath rock overhangs where they could find refuge from the elements but have the benefit of daylight. The inaccessibility of the painted chambers and the lack of detected debris therein suggests that deep caves were penetrated only occasionally. Nobody lived in the painted areas of the cave, as analyses of painted caves' contents have yielded no signs of human habitation beyond the traces of animal-fat lamps and torches used by brief visitors, and some mounds of pigmented-earth left behind. These painted caves were presumably meant to be seen by few human beings under conditions of extreme difficulty and apprehension, as many are entered only by crawling on the belly through a hole in the earth down into dark passages in the earth's womb. These are the archaic conditions that, one may surmise, produced an array of immersive ideals connected to sex and death which became deeply implanted in human immersive instincts and which subsequently became assimilated into Pre-Classical culture (such as the narrative of the mythical Cretan labyrinth in whose belly the deadly Minotaur resided).

Bearing in mind the threat implicit in the hazardous passage that must be made by prehistoric people on entering a painted cave (potentially inhabited by massive carnivores), its remoteness from human habitat, and the expressiveness of the transparently stacked images placed there, I shall suggest that these painted immersive spaces were sites of hypothetical trans-presence. Removed from the illumination of the sun, moon and stars, removed from the daylight realm of accustomed existence, early humans entered into the painted cave's dimness (consequently with maximized retinal dilation) as if returning to the sacred dilated female source of themselves—and simultaneously, to a place of anxious potentiality.

The social function of art within the early formative epoch of human history necessitated, and necessitates, a shared conception of a larger amiable whole, thus the basis of human love and reproduction. With art, people are fastened together by aesthetics into a free-flowing compound-total in the interests of their improved survival, pleasure and replication. As stated, prehistoric art has been discovered at various points inside passages, in niches, and sometimes near cave mouths, but it is in the cave, generally deep within, where prehistoric immersive art attained maximum intensity with its field-of-view encompassing painted murals. These murals will set the precedent for immersive noise art's penchant for constructing overall aesthetic enveloping hyper-totalities which appear continuous by way of their exceeding the normal field-of-view with visual interest.

At first glance, many of the most lavishly adorned murals seem like a noisy chaos of lines and colors. Animals of miscellaneous species emerge at disparate scales and in divergent colors. Also, they are oriented in various directions, even vertically or upside down, some complete, others without heads or extremities. Many are superimposed and thus appear transparent and ephemeral. At some caves, such as Tito Bustillo, though different phases of painting are evident, a corresponding style is used throughout lending it a stylistic consistency typical of the *Gesamtkunstwerk*.[145]

The vast bulk of the remarkably embellished chambers in deep, dark, isolated areas date from the centuries approximately 15,000 BP, the conclusive (but prolonged) phase of the Ice Age. Commonly the walls, which warp and bend overhead (wrapping the immersant in an enveloping total space) are painted and occasionally the floor is also put to use. Always the most immersive salons contain paintings on the walls and, importantly, the ceilings, such as at Altamira, Lascaux and Rouffignac. At Altamira, there are sections of the painted salon little more than one meter (3.28 feet) high, assuring a compressed, close-up, immersive experience. At the Homos de la Peñahe cave, the immersant must lay on his or her back and slither into low hollows to behold drawings.

With prehistoric painted caves, people penetrated deeply into the womb of dark caves to paint and scratch transparent images of untamed animals on every surface of the roughly rounded space, including the floor. As a consequence, we have come to appreciate the sophistication of the noise vision perceptual dynamism which this immersive art utilizes in the transformation of consciousness at a period in time far earlier than the first written words. Hence a feeling for and knowledge of the cave art of Western Europe is essential to a mature awareness of noise aesthetics even though it is conceivable that the majority of readers will not have entered any painted caves, as I have had the privilege of doing. However, as is also the case with noise vision, a personal, experiential understanding of the spatial properties of embellished caves is essential to the development of a comprehensive immersive noise theory, as their enfolding shape and enclosed feel is indispensable to the power of the art. Sadly, these are features which are impossible to convey through flat rectilinear photos.

The 2 kilometer (1.24 mile) long *Niaux* cave system[146] in the French Pyrenees, 5 kilometer (3.1 miles) south-west of Tarascon-sur-Ariége in the department of Midi-Pyréndes in central/southern France is a good place to start on-site explorations of noise vision as this ashen limestone painted cave is owned by the French State and accessible to the public. There is no electrical lighting system inside and if it were not for the torches provided by the guide, the immersant would be in absolute darkness and silence with the exception of faint, reverberating, promiscuous drippings. The ambience is dankly cool, as the cave maintains itself at a habitual temperature of 12°C which markedly contrasts with the tepid air and fervent sunlight left outside. Stagnant water, like a pestiferous dark reflecting pool, covers most of the 50 meter long (164 foot) floor of the first antechamber, which is nearly 30 meters wide (98.4 feet). This chamber leads upward into a high-vaulted, sparsely stalagmited corridor.

One must proceed nimbly and with care so as not to skid on the slimy floor which has been coated by calcite. Pools of tranquil water hinder the path from time to time. There are 700 meters (2,296 feet) between the entrance and the first major change of direction of the long cave, yet the only human markings of the walls are relatively contemporary graffiti, some dating from the Baroque. Deeper inside, some 450 representations await discovery within a complex of chambers, the most celebrated being the *Salon Noir* (Black Salon). This deep chamber is unforgettable because of the distance of the Salon Noir from the cave's entrance. The entry to the Salon Noir is signalled by a smooth stone surface only 1.5 meters high (4.9 feet) from the floor which is scattered with maroon and ebony blots of coloring. Beyond that point, three walls 15 by 20 meters (4.9 by 65.6 feet) are scattered with bestial drawings rendered with jet black contour lines. Moreover, horses, bison and ibex have been etched into the floor at the closure of the space. The subtlety of the 14,000 year old accomplishment is astonishing. For example, in 1974, two engraved bison and an arrow-like sign were detected in a small alcove on the right-hand wall of the Salon Noir despite repeated detailed surveys which began in 1906. The stupendous richness of the paintings and etchings of this chamber construct one of the most spectacular achievements in archaic environmental immersive creation.

In one sense, the Salon Noir is typical of the prehistoric immersive arrangement in that, like most painted caverns, it is entered only after a prolonged and precarious trip, bypassing far more accessible spaces. This journey of course takes committed time, up to as much as three hours at Montespan. Through the moist darkness, prehistoric people passed through unfamiliar spaces (there are no signs of frequent engagement) so as to produce and experience art, even requiring passageway through subterranean lakes at some sites. Also we must remember that caves provided asylum for fierce human predators such as the great prehistoric cave bears, lions, and panthers.[147] Clearly their presence was a dominant factor when we consider that Grotte Chauvet, for example, was found to harbour the remains of around 100 bears. Indeed, certain bear skulls were repositioned to privileged locations in the cave, in one case onto a rock in the center of a circular hall.

Verily, such creatures were *puissant* foes to be feared and assuaged by primordial people, and in this sense caves were not solely sanctuary spaces but also exploratory spaces of fear and sacred trepidation. Indeed Bataille says that the painted cave of Lascaux, for example, was a "place of anguish" and "religious horror".[148] The death risk involved in penetrating many of these openings is attested to by the cave bear-tracks which have been left in the mud floors and along tight trestles. As the risk of death was real, by passing through the mouth of a cave into its admissible swell, the immersant encountered (via dilated retinas) a wide field-of-view artistic phenomenon both sacred and fearful through the prismatic intensity of an adrenaline driven consciousness. Certainly the potential risk encountered, which prehistoric people assumed by traversing such labyrinthine passages, must have been palpable in its production of enzymes. One assumes a *highly emotionally engaging level of alert immersive consciousness* was experienced.

In her book, *Religious Conceptions of the Stone Age and Their Influence on European Thought*, Rachel Levy maintains that such immersive Paleolithic cave feelings have become encoded into subsequent archetypes of beliefs which persist in contouring Euro-American thought and which now, I surmise, continue to move in our regimented grooves of sensibility. If I may conjecture here, perhaps this *attainment of such an adrenalinized cognizant sensation was the point* of the venture, its objective and *raison d'être*, and as such necessitated the descent into the frightful deep pit so as to prepare a stuttering conscious arrival into the adorned chambers rich with depictions of intricately wafting, disembodied forms. If granted, then the opulently noise painted cavern can be said to be *a site of glitch transporting capacity*.

Nerve Noise Visualization in the Grotte de Lascaux

Consequently, prehistoric caves can be seen as places in which *consciousness became self-consciously expanded into a larger field of virtual noise*. This seems well illustrated by the noisy image-shower found in the *Grotte de Lascaux* with its marks of animal transit intermingled with a sense of death and fertility. The most widely known, and arguably the most splendid (looped, as it is in places, with dynamic feral sashes which wind and twist imposingly over its intricate interior shape), is the Grotte de Lascaux[149] located atop an ancient headland in the Périgord, France, which I attained the uncommon privilege of visiting.

As Georges Bataille says, we cannot know the full meaning of Lascaux but we can "sense its maker's desire to impress by stunning our senses".[150] And indeed, in coming into the immersive space of Lascaux, my first impression was of being stunned and disconnected from the norm in favor of a psychic space where sex, art, and death meet in an aesthetic discharge.

Lascaux cave was discovered on the 12th of September in 1940 by four local children and a dog, and shortly thereafter was thought by some to have had served magical imaging functions deemed useful in rousing the psyche in preparation for the hunt. In relationship to immersive noise consciousness, it is necessary to survey what we can ascertain today (given our highly culture-bound predilections) of visualization practices which, it is surmised, were utilized in accordance with the prolifically decorated galleries of Lascaux. We can hypothesize that an ability to visually fashion that which is non-visual (or not yet in existence) by allowing unexpected configurations optically to emerge is essential to life, then as now. This symbolic concentration is a sort of idealized schematization which can be further characterized as a product of *a priori* imagination through which ideas and actions become imaginable. And truly the creative act of visualization is immediately obvious on entering Lascaux's initial salon, as the painters of Lascaux took into full consideration the environmental characteristics and qualities of the physical cavern, first by utilizing both the encasing ceiling and walls, and then by using the physical bulges and bosses of the stone enclosure to meat out the forms of the animals' rumps and bellies.

The painters, evidently, wished to create a total aesthetic ambience that would convey the all-over presence of animals in close proximity to the human visitor (and to each other) as the depicted beasts merge into each other with no respect for the relative size of the different species and with no obvious connection outside of their splendid over-all compositional ornateness. This particular voluptuously painted cave is the most superbly adorned of the prehistoric caves, festooned as it is in a wrap-around overhead garland of overpowering bestiality, with even its ceiling painted (with the use of temporary wooden scaffolding). It is not the oldest[151] nor the largest prehistoric cave, but simply the most artistically achieved and thus the most alluring, from our point of view, as in Lascaux most upper-walls and ceilings are resplendently surfaced with sumptuous immersive paintings depicting the quivering apparitions of semi-transparent animals. Mario Ruspoli characterized these paintings as depicting the "spirits of divine animals".[152] Furthermore, with the *RotundaSalle des Taureaux* (Bull's Chamber), Lascaux holds the distinction of housing the most colossal Paleolithic frieze (with the largest painted figures) known to us and this fact alone merits our rapt immersive attention. One of the Bulls which festoons the cloud-like Rotunda frieze is almost 5.4 meters (18 feet) in length. Others in the same gallery are 3 meters (10 feet), 3.6 meters (12 feet), and 4.2 meters (14 feet) in length, whereas the largest figures at Altamira are only 2.1 meters (7 feet) long and those at Niaux average about .9 meter (3 feet) in length.

Verily, the leitmotiv of the cave is huge groupings of horses in and around large semi-transparent dominating bulls. But what is particularly noteworthy is that this tangle of animal forms exists in a groundless (virtual) atmosphere where the bodies are not anchored to anything suggesting land. Rather, what is suggested is a 360° non-Euclidean space, precisely the arrangement of the ideal range of virtuality. There is no attempt at depicting non-virtual Euclidean ground or defining a landscape, and there are no plants, trees or rocks depicted. Moreover, the dominating figures here are not simply bulls, but rather bull-apparitions, hung with and interposed by a dainty petticoat made up of smaller animals (stags, horses and bison) all organized in crescents and cruseiforms in and around them in interpenetrating and profuse fashion. Furthermore, the mural in the Salle des Taureaux struck me as aesthetically deluxe in its capacity to evoke intelligence through the management of line and its unification of the semi-sculptural with the graphic.

Because the walls of the cavern have been coated with crystallized calcite due to flooding long long ago, the paintings glimmer with a subtle sparkle which is enchanting to noise vision.[153] Thankfully, the congealed calcite served as well as a protective sealing and safeguarding varnish-coat which has kept the paintings' color remarkably fresh and well preserved. This glimmering effect was heightened further when my Ministère de la Culture guide dowsed all the electrical lights (designed to reproduce the tallow lamp originals which burned animal fat with Juniper wicks) and lit a cigarette lighter to better convey an idea of the original visual effect of tallow and burning wicks which provided an unsteady twinkling light (as a candle flame does). At that point the calcite twinkle burst into a full-blown flicker.

More than one hundred burned tallow lamps were discovered inside Lascaux[154] and even if they were all in use at the same time, which is unlikely, one must visualize how faintly dim the light is inside the cave, and how lovely a warmly soft, etiolate-fat incandescence illuminated its walls, and how this flaxen dimness suggests to the mind a semi-dream state, reminiscent, for me, of how invariably exciting it is to go to sleep in an unaccustomed bedroom where the unfamiliar wallpaper and pictures, faultily grasped in the obscurity of night, are only faintly perceptible and thus open to imaginative interpretations.

What is significant to this study of noise vision are the psychic effects produced by the dim seductiveness of the cave's friezes. What I felt when caught up in the supernatural ambience of the space—due to the dim glint of the calculate, the smell of the dank earth, the slightly overhead majestic size and sense of transparent movement of the wrap-around painted beasts which were strewn throughout—was a sense of deliriously (and vicariously) identifying with them, even as they burst over the edges of my visual cone without restraint, and of euphorically running among them as a half-horse/half-man silene (centaur), that jocular classical Greek woodland spirit similar to satyrs (who were half-goat/half-man). This totemistic state of consciousness[155] is what John C. Lilly (1915-2001) calls "species-jumping-thinking".[156] Deleuze/Guattari's term for experiences of this nature is *becoming-animal*. For them, "to become animal is to participate in movement, to stake out the path of escape in all its positivity, to cross a threshold, to reach a continuum of intensities where all forms come undone, as do all the significations, signifiers, and signifieds, to the benefit of an unformed matter of deterritorialized flux, of nonsignifying signs".[157] Along with this experience of feeling imbricated in a becoming-animal panorama[158] by self-fashioning a "map of intensities",[159] I felt enveloped and tangled inside the passion sensation of sacred/sexual noise as the fertile abundance of animal spirits covered and absorbed me in a generalized sense of fertility (a fertility which would help ensure the success of any hunt through a plenitude/excess of the hunted).

This silene sensitivity was particularly acute in the Axial gallery, the gallery that follows the vast Bull's Chamber, as here the cavern tapers to form a more compressed overhead ceiling display. Here one finds a tremendous stag 1.38 meters high (4.6 feet), with an enormous rack of entangled antlers flanked by three horses and an abstract door-like form and rows of dots. Here, particularly, I had this feeling of being included in frolicsome animality. A sense of tragedy was conveyed there too, though, by an apparently wounded and fallen horse which concluded the gallery. Soon, however, my frail humanity gratified me and I felt very remote indeed from the tragic animality of my surroundings, almost as if I were a miniature silene carved out of silver and ivory. As I slipped out of the previously keen feral feeling, I felt the flagrant beasts running over me and exploding me, along with a hundred other things.

Lascaux's friezes, I must assume, had similar psychic/symbolic meaning to those who rendered them and looked upon them, and that they supplied a noise vision framework in which an expanded immersive consciousness could be expressed sociably. Noise vision is present here in various degrees of enfoldments and unfoldments; dividing space up into ostensibly exterior and interior distance has no real significance.

One thing that is unusual about Lascaux is that access to its galleries is far easier than in most other caves (such as at Niaux), with the exception of the gallery called *The Chamber of Felines* which I was not permitted to see due to its remoteness deep within the cave. As previously established, the majority of entrances to prehistoric caves are far from the painted "inner sanctuaries"[160] and require an eventful, hazardous journey which heightens the emotional intensity. Those that are not difficult to reach physically, like Lascaux, start their gallery/sanctuary at the point where light diminishes, creating a transitional emotional and dilational retinal passage adjustment in preparation for a sacred experience (according to Leroi-Gourhan).[161]

The other gallery inaccessible to me was the *Shaft* or *Pit*, which was considered too difficult and dangerous to visit. The Shaft is a 6 meter (20 foot) deep hole, just wide enough for one person to fit in comfortably, halfway along the Passageway toward the Chamber of Felines. It contains the famous scene of the wounded bison who is literally spilling his guts and the bird-headed reclining man with an erection (the sole human-narrative scene in Lascaux). However, I was not allowed to see this.

Nevertheless, just prior to the Shaft/Pit is the *Abside* (Apse), a roundish, semi-spherical, penumbra-like chamber (like those adjacent to Romanesque basiliques) approximately 4.5 meters in diameter (about 5 yards) covered on every wall surface (including the ceiling) with *thousands* of entangled, overlapping, engraved drawings[162] that, on request, I received the additional unique privilege of seeing.

The ceiling of the Apse (which ranges from 1.6 up to 2.7 meters high (about 5.2 to 8.9 feet) as measured from the original floor height) is so completely and richly bedecked with such engravings that it indicates that the prehistoric people who executed them first constructed a scaffold to do so.[163] To me, this indicates that the Apse was an important and sacred part of the cave and indeed Ruspoli calls it the "strongest, most richly symbolic, most mysterious and most sacred" of all the inner spaces making up Lascaux.[164]

Generally the Apse, however, has been ignored by art theoreticians (there is only one widely published scholarly investigation of it per se, by Denis Vialou in Arlette Leroi-Gourhan's *Lascaux Inconnu* even though

Abbé Glory spent several years trying to decipher this inextricable chamber) as nowhere is the eye permitted to linger over any detail (despite holding an immense 2.5 meter engraving (8.2 foot) in its midst). Rather,

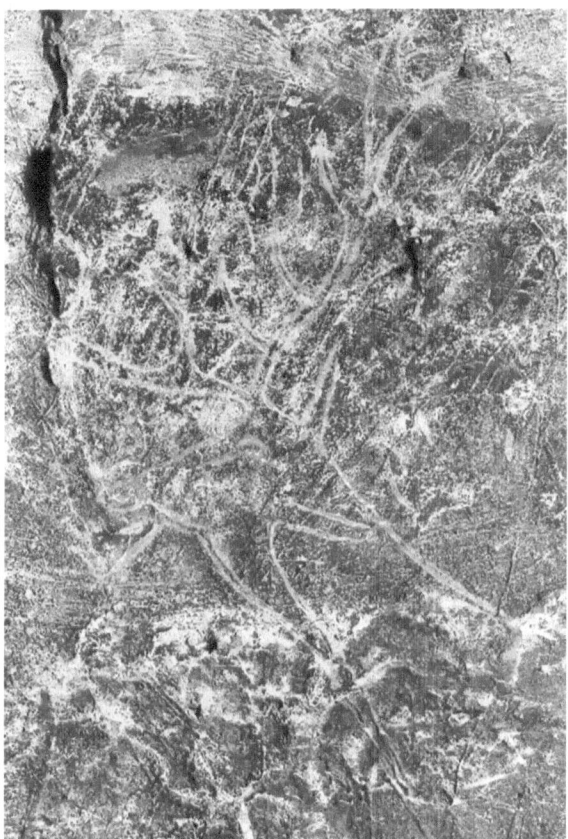

Figure 4 – Enhanced detail image from the Abside
of the Grotte deLascaux, Dordogne (France)

the gaze is urged on by an all-inclusive flood of sublimated optic information in need of visual stamina. Nevertheless, the Apse contains a semi-legible "comprehensive index" of all of the forms of representation found scattered throughout the entire cave, thus making up what Mario Ruspoli calls Lascaux's "véritable corpus" (real body).[165] My assessment, though, is that it is Lascaux's veritable noise vision center.

Describing it, Bataille said that it was one of the most remarkable chambers in the cave but that one is ultimately "disappointed" by it.[166] I was not disappointed, however. Indeed, what pleased and fascinated me about the Apse was precisely its cryptic and foreboding, overall hyper-totalizing, iconographic character granted by its boundless, palimpsest-esque, wall-paper-like image explosion (what Bataille called its *fouillis*) of overlapping, near non-photo-reproducible stockpiled drawings from which, when sustained visual attention is maintained, unexpected configurations visually emerge. Here animals are superimposed in chaotic discourse, some fully and carefully rendered, others unfulfilled and left open to penetration by the environment, all commingled with an "extraordinary confused jumble"[167] of lines including, remarkably, the sole claviform sign in the Périgord and, even more remarkably, Lascaux's only reindeer, an animal that existed plentifully in the period of the adornment of Lascaux. Its extensive use of superimposed multiple-operative optic perception (*optic perception* unifies objects in a spatial continuum) presents the viewer with noise vision par excellence: no single point of reference, no orientation, no top, no bottom, no left, no right, and no separate parts to its whole. Such visual-thought is *homospatial* noise vision, then, as according to Albert Rothenberg in *The Emerging Goddess*, homospatial thought is visual-thought "outside of space or spatiality" which "transcends differentiation".[168] This homospatial quality is deeply suggestive of the non-spatial character of consciousness itself.

As a result of this homospatial noise vision of the Apse, I had the peculiar feeling of being flooded by a cloud-like image cesspool of deep meanings I could not uncode, as if I were in the midst of a model of the Bohm/Pribram universe as implicate pattern. As such, it seemed an imposition onto Paleolithic culture of the very thing that should destabilize it: nihilism. Nihilism, in that it is no longer a matter of heterogeneous figuration, but of scanning a homospatial criss-crossing and oscillating battle scene between interwoven figures, immersed in their ideational ground with which they have merged in a deliberate process of constitutional defigurization. There is no longer any space outside of the figures to define them and, hence, in a mental reversal, space is immersed in the overlapping figures. The nihilistic cancellation at work here, then, seemed to be an attempt to deny the validity of subject/object understanding and to deny that any visual erudition of anything whatsoever is possible, in the interests of omnijective introspection.

Bataille claimed that what was curious about the Apse was that the artists "abandoned their oeuvre to the next to come after them in an ant-like activity", yet "they did not engrave their figures with less conviction or care".[169] Obviously the artists here did not work from a life model but from the overlapping introspective depths of their visual memories. In like manner, the Apse seems to call upon the viewer to construct a mnemonic psychological interpretation of it based on its tightly woven, intricate abundance, that is, its latent excess. But even after introspectively synthesizing the overlapping imploded individual parts into a mnemonic coherent whole, the Apse retained for me a provocative discord and irritation which tantalized my mind further towards a withheld (perhaps forgotten) seemingly encoded signification. But as our subconscious is energized by sustained desire, that which I sensed to be both obscure and overabundant about the Apse merged into a hybrid interpretation that combined conflicting ideas about abundance and nihilism into an *égréore* complex chunk of noise information which I then viewed as a *single meta-nihilistic mega-symbol*.

With this meta-nihilistic mega-symbol's boundlessness, the Apse appeared to me the most sacred of the cave's sacred places. Certainly, easy conceptions of one beautiful being as distinguished from another (in specificity) are denied and an aberrant invalidation takes place where previous concepts of the finite and the infinite implode (as do concepts of the voluminous and the vacuous) into a unified field of multiple-reproductive disembodied existences.

This, then, is a sacred/sexual place of personal intrascoping and transformation (by reason of its creative noise vision and anticipated self-cancellation) as its beautiful representational anti-depictions are neither here nor there but overlap. Clearly, what I am saying about the Apse runs counter to the heart of positivism, a paradigm under which we continue to toil unconsciously, as the positivist ideal is a search for rational, systematic thought where images can be broken down, explored, understood, and explained. Here, in the Apse, we seem to have encountered an irrational systematicism that seems to critique reason, a systematic critique that predates (and in some places overlaps) the modern positivist attitude towards sensation. Here, we are inside a homospatial site of overrunning flux and of hybridization: a place for the rejection of realism and its values (or at least a place to save oneself from the futile and finally unreasonable claims of dogmatic realism and rationalism). The Apse, then, represents a thrusting off of optic and mental boundaries and is thus a complex mirroring of our own fleeting impressions which constitute the movement of our consciousness, the perpetual weaving and unweaving of ourselves. Here, we are not static and have no use for reductive concepts or practices, but are inside a noise space that carries its own nihilistic opposite within itself.

Particularly dense with overlapping noise imagery is the part of the Apse called the *Absidiole*, a small, niche-like hollow (like the semi-spherical small niches that house holy relics attached to the apse in Romanesque basiliques) just in front of the drop into the Pit. Here, the immersant can ostensibly participate in a play of self-tutored multiple-immersion into layers of noise as one stands in the Absidiole of the Apse, which is located inside the groin of the cave, and introspectively view through sublimated excess an explication of the curved inner-logic of immersion into noise itself: *encased and withheld excess*. Assuredly, vision here is no longer the controlling power over animals in nature but, on the contrary, vision itself is engulfed in nature's womb. The motivational force which quickens the Apse, then, seems to be a desire to undermine perpetual vision and replace it with another type of impregnable (immersive noise) vision, or at least to suggest that there may be other types of vision possible. Its nihilistic excess serves the positive function of questioning the validity of the customary appearance of things and to make connective understanding inextricably felt.

Indeed, the basic function of the visual turbulence of the Apse, from the connective perspective, is to precisely shake our conviction that our visual thinking is sound and to hold any such assured convictions, rather, in suspension. Hence it is only routine that formal issues (where consciousness may be said to be self-referential and self-sufficient) would arise over any humanist narrative ethic, as the Apse is more concerned with a recycling of psychological energy than with optically correct astuteness. Hence, freed from representational obligations, dark chaotic powers of consciousness are unleashed via the Apse's repressed excessive exuberance.

When interpreting my immersion into noise in the Apse, we must remember that even the simplest perceptual activity of viewing discrete images utilizes higher-level cognitive activity, as perceiving anything involves description and inference. Indeed perception utilizes a plethora of built-in assumptions and hypotheses as it fills in absent information and draws conclusions based on (but not reducible to) incoming data in terms of part/whole regions and figure/ground relations from which there eventually emerges a preferred percept. Keeping in mind that the human's natural field-of-vision is roughly 120° vertical by 180° horizontal and that the Apse's perceptual-field far exceeds these parameters, the resulting *flooding-over effect* of the Apse (which is significant in creating the immersive noise effect) accounts for some of the visual chicanery experienced here. However, in the Apse, the level of evasive mono-complexity (given the uniform shading in which the one sombre value dominates the complex visual arena) of the *fouillis* also challenges preconceptions of legibility based on our ability to identify and locate figures in their ground, and this made me wonder if the visualization chamber I was in was not perhaps a training spot for the hunters to improve their discerning vision, so as to aid them in visually discovering animals from within their tangled natural camouflage.

But also on scanning the systematic, intricate and perplexing inert spread of the Apse, one cannot but sense that in some way one is looking at a representation of the metaphysics of orgasm and death, and that by absorbing its visual code, one was looking sex/death in the face. To be, or not to be: that is the paradigmatic choice when visualizing form in and out of existence, when examining the elusive alternatives made manifest here. Being, beings, or nothingness: all are tentative conditions of resolution (or forestalled resolution) here; all spout their own ontological/neurological preferences.

In this purging atmosphere of imploded meta-nihilistic sacrilege, spontaneous reflexes only go so far and reflection necessarily takes over in search of an expansive meaning. Yes, nihilistic amanuensis, jubilant noise and catastrophic implosion are here, not only in how this staggering image-dump can be read, but also in terms of how its creation entailed the task of disrespecting the care with which marks achieve representational artistry in an apparent desire to achieve and contemplate radical negation. This scouring of assertive vision must have been deemed necessary precisely here, as in the other galleries, for very often superimposed images respected the marks previous laid down and sensitively incorporated them into the ensuing hybrid super-impositional compositions. By ransacking representational vision in this way, the Apse paradoxically partakes in the category typical of major art (regardless of its marginal standing within the cave and within Prehistory) as it seemingly rejects the figurative tradition in order to reinvent it as entrancing meta—(or supra)—representation. Thus it is major in the way that John Cage's musical composition/non-composition 4'33" is major, that is, in forcing us to astutely consider silence as sound. And, as such, it is a meditation on fullness and emptiness: on the emptiness of fullness and the fullness of emptiness. This is its key noise vision value.

On further reflection, I found the Apse noise encounter to be in rapport with the philosophy of Hegel where he maintains that *our absolute sense is first a pure being identical with non-being.*

Archaeologists are continuously attempting to understand the marks left here from this inaccessible epoch as they analyse its dishevelled iconography in hopes of ascertaining why this tangled impulse was consummated. Most do not see, however, that the Apse defies the common assumption that visual art is associative, that it is based on the human mental capability to make one thing stand for and symbolize another, in agreement with society. The usual assumption is that art-marks on a surface denote content, not just to the mark-maker but to others as well. As an example, the Abbé Henri Breuil (1877-1961) (speaking generally about Lascaux) maintained that some of the mystifying, abstract, geometric marks represented the hunting paraphernalia of traps, snares and weapons, and Leroi-Gourhan placed these abstract marks into a category based upon sexual duality where dots and strokes represented male signs, and ovals, triangles and quadrangles, female. There is mixed agreement on these two interpretations, but all we know for sure about the abstract constitution of the Apse is that its dynamic cluster of representational/anti-representational operations (and the meta-nihilistic/mega-symbol boundlessness which it contains in its kitty) were reworked over the span of many centuries. However, by no means do all of the superimposed figures date from different times, thus their overlapping is not a simplistic function of time nor is it for lack of space. Thus its abstract intentionality assumes a certain degree of lucidity.

The Abbé Glory, who lived in the Lascaux cave for several years while making an inventory of its contents, discovered that in the Apse there are several *re-engraved figures*[170] which is again baffling as it cuts against theories of anti-social resistance to figural thought and places us in the functional realm of *cognitive dissonance,* the psychological term denoting the mental state in which two or more incompatible or contradictory ideas are held to be equally sustainable. Hence the Apse's cognitive dissonance served a virtual function if we remember Brian Massumi's definition of the virtual as "a lived paradox where what are normally opposites coexist, coalesce and connect".[171]

If the Apse functioned as a mnemonic device, or as a site of hegemonic non-being severed from any practical purpose, we shall never know. But it is my hypothesis that the Apse chamber functioned as a cognitive dissonance visualization field and de-focal virtualizing area which adjusted-up the expanding and dilating eye/mind to the awareness of conflicting, non-rational omnijective realities involving sex and death through the use of deeply creative noise visualizations.

We know that most of our cognitive functions and perceptual processes are carried out by the neocortex (the largest part of the human brain) and that the primary visual cortex is the part of the neocortex that receives visual input from the retina. What we can conjecture is that the subterranean aesthetic visualization process at work in the Apse may have been used to feedback optic stimulus to the neocortex in a foreseeing enterprise, an attempt to look into the future, as this process of feedbacking impartial stimulus to the neocortex is roughly the basis for magical gazing. It is imaginable that such a foreseeing enterprise would also be deemed of help in prognosticating the existence and movements of prospective herds of game which would facilitate the success of the hunt, among other things.

To represent the process of this state of looping neocortical stimulus and to fasten a noise cluster of spirit-images on a wall (immersed and hidden among a plethora of others) is in some sense to snare and overpower the image and, ultimately, to have Hegelian power over it.[172] It is curious, however, to note that in the few depictions in Lascaux where animals have been wounded by spears or have fallen, they do not appear to be in pain. Perhaps the seers had found a way of passing into a virtual world beyond the wall by penetrating through the crowded palimpsest-like clutter and joining with the animal's vital spirits.

David Lewis-Williams and Thomas Dowson make the case that, after coming out of a trance, enchanters artistically recreated their visions, both as memory aids for later ritual travels and as portals through which they pass into the spirit world. They view cave markings as powerful ritualistic processes, not just as static pictures, and maintain that the abstract patterns that occur in parallel with the animals found in such prehistoric caves as Lascaux are representations of the phosphenes that accompany the meditative and trance states of the seer's practices, particularly those associated with psychoactive plants. These enchanting practices entailed trance states, it is surmised, which were in some instances produced (in part) by natural chemicals ingested by an enchanter in order to induce a trance for revelatory purposes. Altered states of consciousness induced by hyperventilation, rhythmic movements or psychoactive drugs universally produce *entropic* visual image-fields (a phenomenon derived from the basic structure of the human optic system—anywhere from the eyeball to the visual cortex of the brain—*within vision*). In his book, *Alchemy of Culture*, Richard Rudgley gathered supporting evidence (based on the detailed knowledge of local flora and fungi) from several researchers, that Paleolithic cultures utilized the natural distributions of psychoactive species in their locale as an early feature of their cultural development. *Cannabis sativa* was a known intoxicant in prehistoric Europe and hemp seeds have been found at a variety of Neolithic sites.[173] Trance states, too, were created and augmented by the utilization of hyperventilation and almost always in the context of rhythmic repetitive singing, drumming, dancing and clapping. According to Lewis-Williams/Dowson's adapted three-stage neuropsychological model, people who hallucinate in the later stages often experience a sensation of a vortex or rotating tunnel around them (vortex or tunnel shapes often appear as individuals enter the deepest stage of a trance fostering a sensation of travelling through a passageway). At that point subjects come to inhabit (rather than merely witness) an hallucinatory immersive world.

One may speculate that the Apse served (and/or reflected) such a surrounding process where the self is experienced as capacity rather than existential identity, and where the evaluation of self has been revised from bounded to boundless. Such noise consciousness represents a paradigm shift which relativizes other recognitions of self-consciousness. It is pertinent that, in *A Thousand Plateaus*, Deleuze and Guattari describe this shift towards boundlessness as one's becoming a *body without organs* (BwO) in terms of our self-shifting representational planes emerging out of our field of compositional consistency. According to them, the *body without organs* is an *insubstantial state of connected being beyond representation which concerns pure becomings and nomadic essences.*[174] Deleuze and Guattari go on to say that the *body without organs* "causes intensities to pass; it produces and distributes them in a *spatium* that is itself intensive, lacking extension. It is not space nor is it in space; it is matter that occupies space to a given degree—to the degree corresponding to the intensities produced".[175] According to Brian Massumi, the translator of *A Thousand Plateaus* the *body without organs* is "an endless weaving together of singular states, each of which is an integration of one or more impulses". These impulses form the body's various "erogenous zone(s)" of condensed "vibratory regions", zones of intensity in suspended animation. Hence, the *body without organs* is "the body outside any determinate state, poised for any action in its repertory; this is the body in terms of its potential, or virtuality".[176]

The above scenarios suggest a merging of awareness, first into a more restricted, and then an expanded, intense statement, which is the principle of entering noise art. Thus, it is possible to say that such states of manifestation are distinguished according to the degree to which potentiality is energized through restrictional noise. To apply the noise model to consciousness would suggest that a possible criterion for making qualitative distinctions is the degree to which the potential states of consciousness are unfolded and experienced as a noisy aggregate.

Support for Lewis-Williams/Dowson's visualization account has come from the influential archaeologist Jean Clottes, scientific adviser for prehistoric art at the French Ministère de la Culture. Clottes has joined Lewis-Williams and Dowson in an investigation of their neuropsychological model in an attempt to fill a need for testable theories of why people inconvenienced themselves to such an extent as to create these intensive, highly seductive, immersive spaces. I have taken interest in their work as, from it, we might extract possible immersive art noise intentions and principles from the prehistoric painted and etched inner spaces.

The neuropsychological literature teaches us that trance states proceed in their deepening in stages. Shimmering, incandescent, shifting patterns (referred to in the neuropsychological literature as *entropic phenomena*) have been shown to be produced early on in the trance process when syncretistic noise vision takes on an all-over field-like quality. Resulting entropic form-fields contain grids and lattice designs, dots and flecks, zigzags, curves, and filigrees or thin meandering lines (all apparent in the Apse). In deeper trance states, these fields, depending on the state of mind and cultural penchant of the enchanter, are often, according to Lewis-Williams & Dowson, experienced as a rotating vortex or tunnel that seems as if it was completely sealing off and surrounding the subject in an immersive subjective world. The objective external world is progressively excluded from vision and consideration, and this field of inner enclosure grows ever more florid.

These researchers hypothesized that the art adorning caves, stone shelters and tombs delineate trance-induced immersion into noise stimulated by congesting particular natural molecular arrangements, which produce psychoactive effects in the human brain; these are molecular arrangements which have had a significant cultural history of religious use in inducing visionary, mystical trance states. Accounts of hunter-gatherer and foraging groups include descriptions of enchanters who occasionally conduct rituals that they believe allow them to travel to parallel worlds set out in local belief systems. In these realms, deceased ancestors, deities, and miscellaneous delicate creatures await the enchanter who deals with them in ways intended to meet indispensable communal needs. In preparation for their mysterious interchanges, enchanters typically took steps to instigate trances through isolation in dark places, by frenzied dancing, through rapid breathing, and/or through the ingestion of hallucinogenic plants.

The validity of exploring theories of altered states of consciousness depends on our capacity to overcome that quixoticism which enthrals the mind and takes it no further. That, in turn, depends on the understanding that the subject experiencing an altered state of consciousness remains in principle the same; the consciousness is essentially that of the same person, and the content of consciousness, the ideas and dreams, are those of the same person also, albeit revealed at a heightened level of intensity by the removal of inhibiting agencies and habits of mind. It is on this basis that Walter Benjamin demanded that the revelations of ecstatic visions be made subject to the same criteria of knowledge as those of the sober state, just as *the conventions of conformist ideology must be treated to the same scepticism as one applies to raptures and dreams.*

If one accepts most of what I have said thus far as concerning the alteration of consciousness in the Apse via noise, we may now surmise that this altered consciousness[177] within the Apse would have at least two aspects to it. First, similar to the consciousness shift sometimes experienced when engaging in sex, it is an unleashing liberation and a breaking free from the world's ordinary representational space. This noise domain is one where not only are narrow conceptual territories transcended, but where one also frees oneself from all the desires of security that limit the familiar experience of everyday life. But it is also an enraptured experience which brings noise-fusion-vision into a larger abstract reality, that is, life's covert implicate order where boundaries making up various territories are transcended by our relation to the desire for entirety.

In seeking to understand early immersive aesthetic noise impulses, then, I came away from Lascaux's Apse with a trust in its conjectural goal of serving as a vehicle for *inter-special disembodied connectedness.* Supporting such a noise theory on my part is the so-called sorcerer panel in the cave of Trois Frères, also in the French Pyrenees. Deep underground in a cramped cavern (like the Apse), a rendered half-human/half-animal figure dominates the space. The human/animal figure is staring directly out of the wall (which is unusual for Upper Paleolithic cave art). Just underneath are several heavily engraved panels, a commotion of animal figures with no apparent order or pattern (as again in the Apse). In the midst of this chaos of muddled excess is another human/animal figure and directly in front of this image is a reindeer's hind-legs and rear-end with its female sex prominently displayed. The sacred/sexual immersive (trans-special) potency is palpable.

This proposed explanation for the dark-noise-excess of the Apse cannot be proven, nor, I think, disproven and thus it remains a moot point, however fascinating. Though obviously imbued with meaning, we unfortunately are unlikely ever to know the true meaning or function of the image-space of the Apse (or the other marks of the Magdalenian people for that matter). What I know though, with certainty, is how the immersive noise amplitude of the Apse operated on me, and what it did was collapse the inherited meaning of human image, making into a more inclusive and available sense of excessive ebullition, and a dynamic feeling of wanton sexual climax. Its shrouded noise scattered stirred my desire to seemingly unfold and deliver forth a sanctioned libidinous pathos where forms of salacious creative ferment and levels of self-indulgence are concurrent. From this state of floridity, it might be possible to further define immersive noise states of consciousness as those which *contain a condition in which reality is perceived as consisting of more than that which everyday vision brings to light*. Such immersive noise states bypass discursive counterintuitive processes and confer a greater scope to vision and therefore an enhanced and expanded unanimity.

Bolstering this contention is the fact that, before leaving the Apse, I had looked around down the Passageway and into a portion of the Salle des Taureaux and I recall these chambers taking on the character of a moist orifice. At that point I felt like a naughty ravisher about to act out some unfathomable, risqué, multi-genus sexual act, as if I was emancipated to ford my human anthropocentric sexual frontiers and burst out of my specific species identity and into that of a bull, horse, peacock or peccadillo just as I have frequently imagined myself doing when engaged in sexual union. It is this sense of inhabiting a new corporeality in obbligato that is entirely unnatural, preposterous, and variegated which, as we shall see, holds importance when uncovering the idealized desires and onanastic qualities of the immersive noise art experience.

What additionally fascinates is that this fine jumble of delicate lines, some beautifully representational while others not, corresponded to the prolonged series of greyish drawing with which I began my career as an artist some twenty plus years ago: drawings which had partially been conceived of as a shadow of our nervous system's meshed neural signals.

Thus the Apse seemed an idealized shred from my own memory and I nearly felt that from the ceiling angelic divinities would pelt garlands of roses down on me. We should note that it is common to find prehistoric stones of various sizes that were incised with a jumble of overlapping animal drawings in no apparent order, piled on-top of one another to the point of illegibility.[178] We can say with assurance that the Apse's brimful-room noise style is almost unprecedented, save for certain panels in Les Trois Frères and at the cave of Combarelles, a nearby Périgord cavern which I subsequently visited the next day.

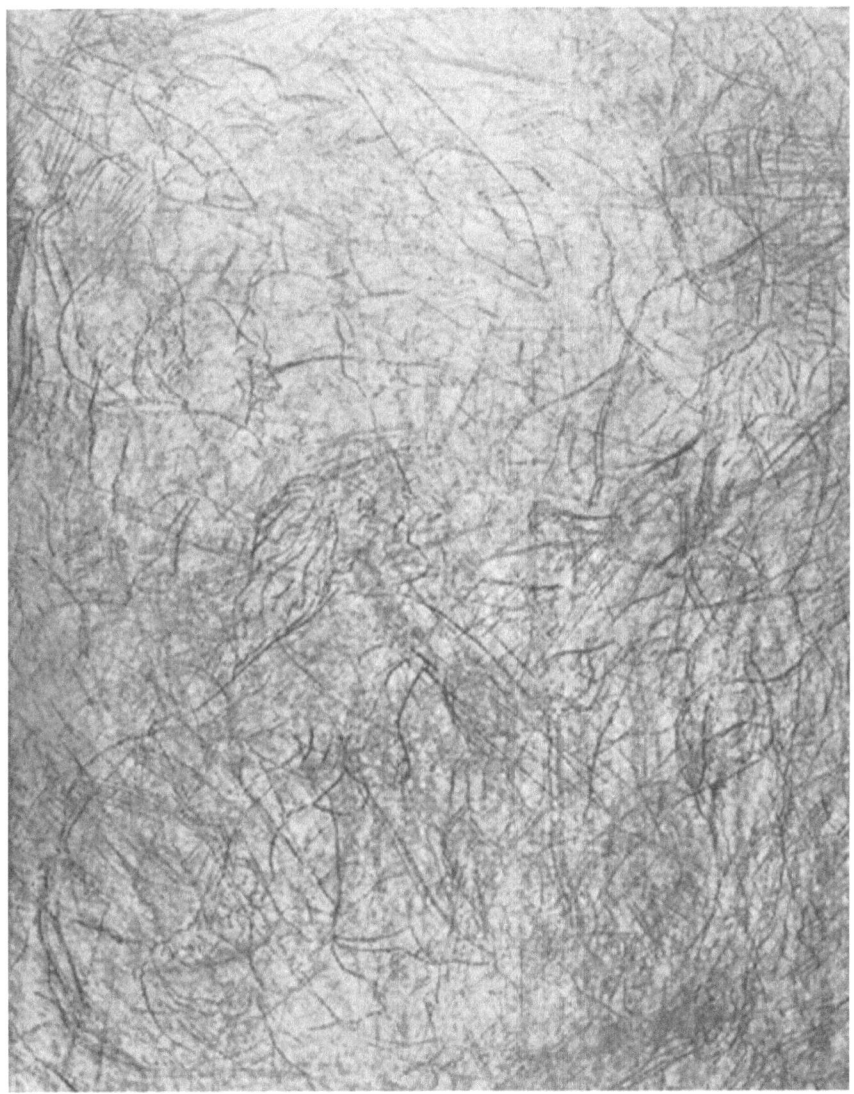

Figure 5 – *Gods of Politics*, 1984, 14x11" graphite on paper, Joseph Nechvatal

On exiting the cave of Lascaux, the sense of psychic openness was striking as one returns and runs into the light, one's eyes reconstricting as one passes through the sparsely wooded area and emerges into homogeneous light on top of the hill with a magnificent vista at one's feet. It was there I spent the night in an auberge in preparation for a visit to Combarelles.

The cave of Combarelles, like the Apse, contains an enormously doleful pile-up of almost imperceptibly engraved drawings deep, deep within the once almost inaccessible wet belly. A prolonged walk inside the cave preceded any encounter with the art but, once encountered, like in the Apse, depicted forms start snowballing and overfeeding on themselves. Here, too, our visual-mental system self-devours the assumed reality principle, ultimately causing its downfall by absorbing realistic representation into a homospatial noisy dissolution of form. Like the Apse, it too is colorlessly elaborate, heady, and intricately composed, but here I felt neither ravished nor aroused nor stretched by the hyper-fastidiousness of the obscure excess, but rumpled and crushed beneath the cave's monotonous dark and inaccessible logic. Indeed here, as in the Apse of Lascaux, representation was problematic and the normal linear depiction of figurative assurance failed in favor of a multi-linear non-sequential processing. Certainly, the etched walls did not have one singular classical point of view or a fixed position from which it depicted being, and it, too, operated on the dynamic of a supra/meta-dataload. But this operation was never mitigated by other colors of thought which might have allowed Combarelles to transcend the limitations of its own pictorial assumptions via a critique of them, as Lascaux had managed to do.

What the open-endedness of the piled-up, noisy, disembodied fabula at Combarelles suggested to me was the collective abstraction of the production and distribution of every possible representation, along with the super-human desire for existing pluralisticly in many orbs simultaneously.[179] When I thought of the hyper-connectivity of its indistinct veneer of interlaced lines, I saw Combarelles as a meta-idea cove which functioned by criticizing the discourse of traditional understanding through measuring the distance and difference[180] to which coherence goes, and indicating from whence it has come: the complicated blurriness of noise.

Nymphaea Nerve Noise

Examined through the tradition of communicative symbolic interaction, immersion into noise's prevalent territorializing/deterritorializing configuration thus far appears to me to be roughly the *inscribed parabolic space* as we saw in the Apse of Lascaux. And, as such, noise art begins to create a cultural domain which is half illusionary and half real, just as any symbol is.

Rounded noise order seems an attempt to encircle vast shapeless infinity into a symbolically distinct scope and location through parabolic configuration. Hence, immersive noise consciousness seems thus far to be primarily a function of a desire to create a convincing illusion of non-self-containment through a semi-enclosed noise space which heralds the sanctum of the tribal magic circle, the circle which interpiercingly severs a space of sanctity from the profane. According to Nigel Pennick, the circle is one of the most ancient symbols used by humanity and is seen through the history of humanity as the embodiment of the universal whole, representing the perfect totality of the macrocosm.[181] It symbolizes the perfection of totality in that the circle is a geometric figure formed with one line with no beginning or end.

The central spot of the ancient symbolic immersive circle is the *omphalos*, the pivotal, still, capacity-point within the sacred circle. Inside the sacred immersive circle, the outside world is dominated and indeed defined by the omphalos' psychological protectoratship. The conceiving mentality behind the omphalos was that it marked the fixed point of the earth around which the spherical spiritual heavens whirled. Thus it represented a central place which remained steady and enduring while all else moved about it.

Today we know that the earth rotates on its axis once a day, and that it revolves around the sun once a year. In early times, however, astronomy was based on an ideal geocentric cosmology according to which the earth was fixed and immovable. The earth was conceived as being at the center of the universe and everything spun around it. In this cosmology, the universe itself was imagined as being bounded by a great sphere to which the stars, arranged in the various constellations, were attached. So while we today understand that the earth rotates on its axis once every day, in antiquity it was believed instead that once a day the great sphere of the stars rotated around the earth. As it spun, the cosmic sphere was believed to carry the sun along with it, resulting in the apparent movement of the sun around the earth once a day.

The omphalos' quintessence may have been only a scant central fire within a circular placement of stones on the ground which carved out the immersive space of emotional sanctity. However, an interpretation of this hoop of stones with centered still-point may be quickly conceived in terms of recognizing a point of view within the cyclical arrangements to the surrounding cosmos, as we see with the omphalos' evolution into the classical Greek maypole. A circle with a marked center and circular design elements emanating out from the central point is almost universally found in the world and it forms the basis of the floral rosette, one of the oldest and most widespread of ornamental designs.[182]

Accordingly, since its Mediterranean origin, western philosophy has fundamentally presented itself as a theory of the omphalos. And with this idea of the fixed, sacred, central spot we see the nucleus of the city/state, as the sacred staff of the seer (which was used to inscribe the perimeter of the sacred round circle) turned into the phallic obelisk (rather than the female pudendum) and begins marking the convex power point around which all is organized.

We shall quickly see in this and the next section how the sacred psychic circle (constructed around a central omphalos) connects to the sanctuary of the encircled sacred grove which itself connects to the origins of art in the West and to the maturation of the city/state. Thus far we have established that a parabolic immersive noise site is interiorly and conceptually encircling in aesthetic immersive sites, in order to enable the swallowed/semi-assimilated subject no avenue of self-protective flight from its excess of signification. What we have seen with the pudendum-like prehistoric embellished cave is that the prehistory of immersion into noise is primarily a history of assertively embellished aesthetic space in service of the virtual, the peripheral and the mercurial. It is for this reason that we will turn our attention now to certain aspects of the nymph myth continuum which makes up the enchanted *nymphaea* garden grotto[183] legacy, for the phenomenological awareness which such a lissom simulacra provides this discourse shall be serviceable in flushing out the extensive meaning of immersive expectations.

Nymphaea is the Roman term used to describe temple fountain-shelters consecrated to the nymphs which were based on simple Agora grotto water spots. A *nymphaeum*, under the Romans, became a formal temple dedicated to the cult of a nymph. This temple often related to the source of a stream, but because these structures were based on the Greek natural grotto grove (with spring), the term later became applicable to both artificial fountain grottoes and to monumental public fountains.[184]

Descended from classical and eastern Hellenistic prototypes, grottoes proliferated in the late 1st century BC and spread further during the Imperial era when they became a common feature in the gardens of wealthy landholders. A rigorous definition of the term *nymphaea* would limit its designation to sacred semi-enterable edifices that served as sanctuaries of the nymphs, and this is the sense which I am using the term here. Another important distinction to maintain, however, is that between the public nymphaeum and the private nymphaeum. Two principal types are evidenced in both cases: the rustic grotto niche, in imitation of the Arcadian cavern, and the architectural fountain-temple type (for example the, now chiefly collapsed, immense Nymphaeum Hortorum Licinianorum, or the extant Castell dell'Acqua Marcia, both in Rome). In private hands, the interior nymphaeum was often located within an architectural apse or in a large niche comparable to the *cavea* of the theater. The apse/nymphaeum constitutes the primary feature of the House of the Great Fountain in Pompeii, for example.[185]

Clearly, entry into art noise space is not so much an entrance into earthly expanse as it is a representational passage into non-space *ad infinitum*. Noise is the space of access/excess we know as part of the instantaneous computer communication world. Noise is also encoded into the nymphaea origins of the garden grotto with its legacy of immersive exaltation of the feminine and its endorsement of sumptuous love. Thus, it is upon the garden grotto's roots as a sacred/sexual nymphaeum grove (based on the sacred omphalos-pudendum) where we shall begin to build upon the previous section's noise recognitions by continuing to trace the outgrowth of noise culture as detected in arcane archaeological sources and philos-theological traditions, both of which are open to interpretation of course. What is stimulating about the noise nymph and the omphalos-pudendumic nymphaea tradition for our purposes is its usefulness in tracing the cast-around 360° ideal aspect of noise within an enclosed, or partially enclosed, container.

The grotto can be seen to embody the bucolic or the idyllic, the sacred or the profane, the mythological or the prescient, and/or simply be eloquently ornate. In a sense, it is the space of anti-noise. The grotto's space is the space of tranquillity, coupling, solitude, seclusion, obscurity, and cool pathos; but most significantly it is traditionally a metaphorical space symbolizing the human vector within the unbroken universal matrix.[186] But any metaphorical topos for the universe must be in its very constitution indeterminate, noisy, complex, unified and unsatisfactory in its denotation. The nymphaeum is that too, as its various definitions and types are capriciously broad while all sharing an accordant meaning.

We can take the labyrinth as a symbol of immersion itself, as the entire point of a labyrinth lies in getting lost and searching about,[187] along with the self-discovery encountered through the search. That and their necessarily willed abandonment, all of which is salient to noise consciousness. Hence, labyrinthine understanding offers an understanding of works of noise art in that it grants us experience by penetrating space/time and, in a sense, secures that space/time for us.

The labyrinth is a cultural noise space blending both landscape and architecture into an intricate search. In ancient times, when pregnant animal carcasses were cut open and disembowelled in preparation for consumption, there inevitably would be a great outpouring of the winding intestinal tract mixed up with the foetus. Not knowing anatomy as we do, it is supposed that primordial people took the winding intestines to be the birth canal. As a result these beliefs became part of Pagan lore.

The earliest surviving labyrinths, all of classical seven-ring design, are rock carvings and graffiti and patterns on coins, seals and ceramic vessels, rather than full scale forms that could be walked through or upon. Full-sized labyrinths were too vulnerable to survive thousands of years against the combination of neglect, erosion and overgrowth. Early surviving labyrinth designs are found carved on part of an ancient dolmen at Padugula, Nilgiri Hills, in southern India which dates back to 11,000 BC, on a 1,300 BC ceramic vessel found in Syria, and on a 1,200 BC inscribed clay tablet found at Pylos, Peleponnesos, Greece. The labyrinth carving found inside the *Tomba del Labirinto*, a Neolithic tomb[188] at Luzzanas, Sardinia, could conceivably date to 2,500 BC if it is contemporary with the tomb, but later burials make this uncertain. There are at least five labyrinths carved into rock faces above the town of Capo di Ponte, Val Camonica, in northern Italy, ascribed to the Late Bronze Age or Early Iron Age (1,000-500 BC).

Crete, considered the place of origin of all of the Greek Gods and Goddesses, was a highly developed Pagan civilization before its volcanic destruction in circa 1400 BC, with active trade routes to and from Egypt and other lands in the Mediterranean. Various Cretan coins between 43

BC and 67 BC bore the classical seven-ring labyrinth design, both in square and circular forms. This classical labyrinth design is believed to have originated with the Cretan parable of Theseus and the Minotaur. According to Greek mythology, King Minos of Crete had a craftsman (Daedalus) construct the labyrinth in order to conceal the Minotaur; the half-bull/half-human progeny of Minos's wife Pasiphae and a bull-Zeus. Queen Paisiphae, evidently sexually unsatisfied by King Minos, had ordered the inventor Daedalus to construct a convincing full-size model of a cow in which she could conceal herself, exposing only her vagina. Zeus, greatest of the Gods (who was born inside Ide004 Cave on the island of Crete) descended in the form of a bull and mounted and impregnated her, resulting in the birth of the half-man/half-beast Minotaur.

There are several variations of the legend of Theseus and the Minotaur, but the main story is certain. Crete had won a victory over Athens and as a cruel tribute required that every nine years seven young men and seven maidens should be sent to Crete to be devoured by the Minotaur, who was now confined in the labyrinth. The fourteen victims were chosen by lot, bringing terror to every family in Athens whenever the tribute became due. Finally, Theseus, son of King Aegeus, volunteered to resolve the matter by slaying the Minotaur. Aided by a ball of golden thread provided by the King's daughter Ariadne, Theseus entered the labyrinth, slew the Minotaur and exited the complex space by following the golden thread he had unravelled on his arrival, thus finding his way out and ending the cruel tribute.

This myth was widely known, as Zeus is a central figure in Greek mythology and, hence, became familiar in subsequent Roman culture. At Pompeii, where I visited, there was a square shaped seven-ring labyrinth scratched onto a crimson painted pillar in the House of Lucretius some time before the city was destroyed by the eruption of Vesuvius in AD 79. It has around it the cryptic words *Labyrinthus, hic habitat Minotaurus*. This demonstrates that the Romans were well aware of the Greek Minotaur's sinuous labyrinth.

Although not in the classical design, the labyrinth motif was used in mosaic pavements throughout the Roman Empire and these are the oldest surviving full-sized labyrinths. A significant variation on the classical labyrinth design is the addition of a second entrance (or exit), so that a procession can enter by one entrance, reach the center, and then emerge by a short exit without turning around. The design is still essentially unicursal, however. The most enduring Roman labyrinths were built in mosaic as such mazes. Other Roman mazes are complicated networks of paths, like a labyrinth. However, unlike a labyrinth, they have multiple openings and possible directions (not just one as in a labyrinth) which succeed.

The medium of mosaic offered much in the way of permanency to labyrinth and maze design. As well as being durable, many Roman mosaics were shielded from subsequent erosion by the collapse of the very buildings they once adorned, thus many examples have survived. Roman mosaic mazes consisted generally of a rectangular grid for most of the area which they filled, using the central area for pictorial illustration. Normally square and the size of a room, the most popular subject was the slaying of the Minotaur, but some Roman labyrinths simply portrayed the Minotaur, or other half-human/half-animal creatures such as centaurs. Eventually maze patterns were incorporated into the floors of some Catholic churches and cathedrals (less the Minotaur) such as in the nave of Chartres Cathedral which contains a majestic maze 9 meters (30 feet) in diameter to which penitent Christians peregrinated on their knees.

In my noise vision view, the earth is a kind of wild vibrational arena in which one omnijectively experiences the pleasures of the flesh while being cognizant of the fact that one is an expanding noise projection immersed in an amplifying orchestration. The effectiveness of such a noise aesthetic realization depends upon one's advancements in the area of intellectual and emotional conceptions rooted in noise. Fortunately, the pudendum-based grotto is possibly a site *par excellence* in which to scrutinize this obviously thorny province of voluptuous noise vision.

To concentrate on the grotto is to summon all that was said concerning the archaic painted cave. Like in the treated cave, the art of the grotto uses (and then surpasses) nature to concoct an apparatus deemed suitable for shaping cognitive-vision/consciousness along the lines of the attributes of the omnijective expanding universe by modelling dilating connectivity in miniature. The discovery in the late-1920s by American astronomer Edwin Hubble (1889-1953) that the universe was expanding implies remarkable things for the immersive space of the arcane grotto, as, like the painted cave, the grotto is a miniature zone of expanding liminality and cognitive crossing. It is a space of escape from the world of naive naturalism (for example, that proposed by the Italian theologian/ philosopher Thomas Aquinas (1225-1274)) and a zone of entry into the fluid, rhizomatic, and elfin world of connectivism where the spatial restrictions of conventional realism (think of the paintings of Jean-François Millet (1814-1875), Thomas Eakins (1844-1916) or Winslow Homer (1836-1910)) need not apply, even while biological nature remains the grotto's starting point. Withdrawn into this zone of fay interchange, the immersant joins consciousness, not so much with the world outside, but with the classical Arcadian inner world of unconscious preterhuman existence, with its mantric cerulean rites of birth, pubescent passage, coupling, incantation and death.

Porphyry (circa AD 233-303), the neo-Platonic and Neo-Pagan author of *De Antro Nympharum*,[189] tells us in the French translation that, even in the earliest times, certain caves and natural grottoes were consecrated to the gods and goddesses, long before temples were conceived of and built (citing the cave of Lycean Pan in Arcadia, among others).[190] By way of preparation for the grotto, archaeological evidence has indicated that there are traces of a 15th century BC Egyptian sacred garden grove in the temple complex at Karnak. I visited that old garden spot, which is tucked away deep inside the complex behind the sacred sanctuary temple of Amun (the hidden one), and found it barren but most immersively suggestive with its inner placement and diminutive scale.

It was from the Assyrian civilization in northern Mesopotamia that we find *sacred groves* within which modest shrines were contrived for supplication. Moreover, archaeological evidence shows that some Mesopotamian structures had pits positioned into their rooftops which were planted with a variety of sprouted ferns and flowers that constituted a minute garden site for contemplation connected with the cult of Tammuz and/or Dummuzi. These sacred cults were later imported into Greece where similar sacred groves were claimed in the wild, but now based on dissimilar female divinities called *nymphs*.[191] The grotto's noise poetics can particularly be traced to the coves dotting the coasts of Greece, such as the dazzling caverns in the Peloponnesus along the bay of Diros[192] or the one I swam in daily in Siros for a week in 2008.

Exactly where the Greek concept of the abounding sacred sexual nymph stemmed from is not known. I assume it is a descendent of the cult of Hathor from North Africa, but why this concept arose in North Africa, we do not know.

Ostensibly, in ceremonial observance of this long fertile tradition, there emerged the previously mentioned pudendumic nymphaeum, an ancient Greek secluded area dedicated to the nymphs which typically included an extemporaneous grotto with waterfall or spring, nestled in a grove of trees (or sea cove) with a central devotional arena. This reminds us again of the Greek *temenos*, the spot removed from the common land, dedicated, in this case, to nymph Goddesses. The pudendum provides the nymph worshiper a full or semi-encircled sacred immersive space in which to enter into communications with the nymphs, for example with Syrinx, an Arcadian nymph who turned herself into a reed to escape the advances of the shepherd God Pan.[193] Pan, who lived in caves, was son of the nymph Penelope and is thought of as the God of fertility and unbridled male sexuality, known for engaging in sexual activity with various nymphs in the form of a goat. No cave dedicated to Pan and the nymphs is more renown than the Corcyrian Cave on Mount Parnassus which is celebrated as the site of numerous Bacchic orgies.[194] Yet Pan is not to be confused with *satyrs*, who were Greek woodland spirits. Satyrs had a human upper body and the lower body of a goat and were generally depicted as having dishevelled hair with goat horns and ears, and with an exacerbated erect penis (ithyphallic). In early Greek art they were portrayed as offensive in appearance, but later they were represented as being handsome and sexy. Greek vases occasionally depict post-coital sleeping or sexually active nymphs such as Thetis (who attempted to make Achilles, her son, invulnerable by dipping him in the waters of the river Styx).

Few places testify more vividly to the development of the grotto than the cavern rich Bay of Naples. Insofar as the sea-based nymphaeum was incorporated by Roman culture into Italian gardens in the form of small grottoes with fountains or limpid pools of water, it advanced an eventually widespread European garden tradition (as Italy set the model for all early sophisticated European gardens). Grottoes in the Italian style generally present a pastoral, semi-nude nymph from Pagan fables (frequently Venus, the Roman adaptation of the Greek Hathor-based Goddess of love and beauty Aphrodite, whose myths she took over) tucked into a niche and accompanied by ferns and spouting or bubbling water. Venus, it must be remembered, was the Roman Goddess of love, originally associated with the biological fecundity of vegetal gardens. Amor, Roman God of love (the equivalent to the Greek Eros) was the son of Venus. Venus's cultural importance rose with the political fortunes of the clan of Julius Caesar (circa 100-44 BC) who claimed descent from Venus via Aeneas and Julia. Indeed Caesar instituted the cult of Venus and proclaimed her the Goddess of marriage and motherhood, Venus Genetrix, under which name he constructed a temple at the Forum in her honor. Her festival, Veneralia, is celebrated on April 1st. Most people today know of her from the 2nd century BC Hellenistic sculpture *Venus de Milo*, which was purchased by France and brought to the Musée du Louvre after her discovery in 1820 on the island of Melos or from Tiziano Vecellio Titian's (1485-1576) 1519 painting *Worship of Venus* at the Museo del Prado.

Dionysos is Back

What is also important to an immersive noise vision theory is that classical Greek chorus drew its associative power from the character of the thrice sacred impulse of the ancient agricultural Springtime Spree of drinking, planting and encountering ghosts. The Greek chorus is a remnant left over from the above mentioned ritual forms, in which all male community members participated freely and for which Jane Harrison uses the terminology *dromenon* (the thing done).[195] This ritual-action–turned-communicative-presentation is consistent with what Emmanuel Levinas (1906-1995), in *Totality and Infinity*, says is the basis of the social relations: free gift-giving (so that referents can be held in joint to crystallize their communicative reciprocity).[196] But what is most significant to noise theory is the circular *orchestra*, the space on which the chorus noisily sang and danced. The relationship of that circle to eventual spectators shall illuminate just how noise art arose out of ecstatic ritual. One must remember that the tragic dramas of the poets Aeschylus (525-456 BC), Sophocles (495-406 BC), and even Euripides (480-406 BC), it is thought, were played not upon the theater stage but within the circular orchestra, one which marked out the sacred patch of the gods and goddesses.

Originally, tragic drama in Greece consisted of a single actor and a large chorus, suggesting that tragic drama began as a choral celebration in memory of a dead hero (a replacement for the fawn or goat) in which someone, probably the leader of the chorus, at some point began to act out the exploits of the person being celebrated (after being symbolically eaten). In roughly 550 BC, the Greek Classical age began with Aeschylus, a notable participant in Athens' major dramatic competition, the Great Dionysia (a part of the festival of Dionysos). Aeschylus's influence on the development of tragedy was fundamental in that previously Greek drama was limited to this one actor and the chorus. Aristotle tells us that Aeschylus was the first to introduce a second actor. Aeschylus's tragedic production work was followed by that of Sophocles, work typified by tragic reasoned thought and polished phrasing. Aristotle tells us that Sophocles was the first to introduce a third actor into the tragedy. Sophocles' work was followed by Euripides, the tragic poet who is most responsible for severing the chorus from the action of the play. Aristotle tells us that, by Euripides' time, it is clear that the number of main actors has increased and the importance of the chorus decreased. Euripides' work also interests us in that he was predominantly an investigator of intense *viractual* [197] conceptions. A relevant example of Euripides' work, which was brought to my attention for its noise importance by Miranda Aldhouse Green, was the play *The Bacchae*, the last and greatest work of Euripides. Through briefly looking at this play I hope to show something of the noise nature of Greek tragic dramas as they were experienced by the Athenians at the Great Festival of Tragic Drama, an annual religious festival in honor of the God Dionysius.

The Bacchae, which is narrated by the chorus (consisting in this case of female worshipers—played by masked men—of Dionysius called *Bacchae*, a name derived from *Bacchus*, the Lydian name for Dionysius) tells the story of Dionysius, the Greek God of wine, revelry and of nature in all of its organic and bestial prodigality. The Bacchae refers to a group of *maenads* caught in Dionysius's Bacchic frenzy, whipped up by the exacerbating attractive enchantments of Dionysius.

In *The Bacchae*, Dionysian ritual is consistently connected with exultation and liberation as the chorus sings of the raptures of Dionysian bliss. Such Dionysian worship was only one of the mystery cults that flourished in ancient Greece, however, the most widely known being Eleusis and the Eleusinian Mysteries. The word *mystery* here refers to the fact that these cults required that their rites be kept secret from outsiders. Most scholars believe, on the basis of testimony from Clement of Alexandria and Tertullian, that the Greek Mysteries were comprised of three main components: the *deiknymena* (things shown), the *legomena* (things said), and the *dromena* (things done).

In the play, by enflaming the Bacchae, Dionysius deliberately rouses the anger of the disrespectful but authoritative, youthful King of Thebes, Pentheus, who vows to put a halt to the Dionysian orgies (the Greeks called the rites of mystery cults *orgia* (i.e., orgies)). Enraged by Pentheus's refusal to accept his ecstatic authority, Dionysius whips the women of Thebes into a deranged and furious delirium as to castigate Pentheus's impertinence.

The play's course covers Pentheus's attempts to dissolve the tenacity of Dionysius's necromancy and his eventual humiliating demise at the hands of Dionysius when the disguised Dionysius shrewdly causes Pentheus to challenge (and ultimately relent on) the full force of his powers. By so doing, he compels the King towards his own destruction, notwithstanding efforts made by his grandfather, Cadmus, and an eyeless augur, Tiresias, to discourage Pentheus from his agenda. Dionysius deludes Pentheus by making the King see him as a bull, to think that the palace was in flames, and believe that a phantom Dionysius, which the King was trying to stab, was the God himself. Dionysius appears at the end of a tragedy as *a deus ex machina* (God from the machine).

The orchestra in which this work, and others, were first played consisted merely of a circular plot beaten flat and sometimes edged by a stone periphery. This is perhaps best seen today at the Epidaurus Theater where the circle is now surrounded by a *theatron* (the spectators place) which was subsequently added on. The theater, for the Greeks, was simply *the place of seeing*, (where the spectators sat) and the *scene* (or *skene*) was a hut or tent in which the actors dressed.[198] The central focal point of the whole was the *orchestra*, the circular dancing/playing/singing arena where the chorus of men performed their tragic dithyramb. It is from this active arena where the ideal (an ideal ironically for both the totality of the Gesamtkunstwerk and for non-art) of the non-differentiation between artist and non-artist, between art and life, between noise and music, between various art disciplines, and between the final work of art and the spectators, originated in the West. All these impulses stem from the group revelry taking place in a noisy sacred circle which sprung from the hoary shrine. It is this relationship between the space of the chorus and the space of the spectator where we can observe, with the shifts of time, the emergence of art noise from its roots in participatory ritual—the move from *dromenon* to drama.

The space is circular because its quintessence is the previously mentioned circular arrangement of stones on the ground. It procured a sense of fervent sanctity in which the undifferentiating dance-rite revolved around some sacred/sexual focal point at the circle's center. As previously outlined, this centering point (omphalos) represented the centered place where heaven joined with the earth and where communications with the gods and goddesses were possible. It is from this metaphysical hoop's omphalos that occult noise perception generally looked inward at cocooned inner space and outward towards an expanding immersive space of the vast cosmos. At first this point was marked by bundled stalks of reaped oats that sat in the center of the circle and only later became a stylized male phallus or female pudendum or the figure of a *homo erectus* god or goddess, and then still later their extra-representational maypole or altar. This sacred centring point of encircling immersive space reflected the belief of the centered place of the community member in the cosmos.

In the circular space of the proto-orchestra circle, the entire licit Greek male society would gather and ardently rotate around the omphalos cum stave.[199] There is no division at first between actor and spectator, as all Greek men participated in the dance-worship with its consolidated emotion. In all respects, the amphitheater seating, which we know well today, developed when the Greeks moved the omphalos-based sacred orchestra circle up against the side of a slopping hill so that those excluded, but watching (the uninitiated, the women and the children), would have an unobstructed view of the Dionysian festival. The Theater of Dionysos at the Acropolis is a chief example.

With this new arrangement, more and more uninitiated people would gather to watch the ceremony and it is precisely at this period where the Dionysian ritual, the thing actually done, turns into the abstraction of art—and into show. *Thus a bulk of western art as it has been conceived for about 2,400 years begins with the demise of immersive noisy participation and the advent of passive contemplation* through the watching of something prepared worthy of attending. Now the noise eye and ear has been removed from the action of the rite and separated from the whole and placed at rest, aloof and detached through distance by the mounting stone seats which semi-circle the spherical omphalos-based orchestra pit.

What an emphasis on aesthetic immersion into noise does, is to place us back into a ritual position by dragging art down into the felt 360° noise-perspective of the enthusiastic and participatory (if we fight to overcome cultural impediments).

CHAPTER 3

Signal-to-Noise Eye

Anti-Noise Vision of Linear Trompe l'oeil Perspective

The Renaissance is linked with the rediscovery of classical culture. Consequently, with the revived interest in antiquity came a new repertoire of Pagan subjects for art and this interest in Pagan matters facilitated the re-emergence of the grotto within the Renaissance pleasure garden.

Concurrent with the Renaissance pleasure grotto, however, is what came to dominate the Italian Quattrocento: the development of rational, linear point-perspective, the technical perspective rendering of a scene from one fixed and tapered eye-point.[200] As Robert Romanyshyn describes in his book, *Technology as Symptom and Dream*, linear perspective vision "achieves a kind of geometrization of the space of the world, and within that space we become observers of a world which has become an object of observation".[201] This "objective" rendering, with its emphasis on the horizon-line and vanishing point, formed the pictorial ideals for painting and drawing, of course, but also it formed them for the Italian, French and German Renaissance garden itself. Walking into a Renaissance garden, such as the Ville d'Este, Château de Champs at Marne-la-Vallée, the Château de St. Germain-en-Laye, or any of the sumptuous gardens of the châteauxs of the Loire Valley (such as at Chambord, Blois, or Azay-le-Rideau), one has little question about the key values they amplified: human reason and power justified by a godly transcendence in reunion with classical antiquity. These values are articulated by what Romanyshyn sees as the central function of linear perspective, its "celebration of the eye of distance" which becomes elevated into a cognitive methodology.[202]

By looking over the orderly formal garden through a tightly focused perspective, a sense of visceral but distant scopic power becomes evident.

Samuel Edgerton, in his *The Renaissance Discovery of Linear Perspective*, characterizes this focus as "a means for organizing the visible world itself into a geometric composition, structured on evenly spaced grid coordinates".[203] Unfortunately, with geometric organization any sense of intimate sacred/aesthetic noise vision/contemplation is lost in favor of the metaphysics of scopic power. The strict formal gardens of the Renaissance represent a major step, then, in the domineering, framing and rectilinear boxing-in of noise vision, and a repression of the scope of immersive propensity which we have been exploring thus far. Worse, according to William Ivins, this repressing framing tendency moved "from its discovery or invention as a quasi mechanical procedure to a logical scheme or grammar of thought".[204] Moreover, according to Norman Bryson in his *Vision and Painting: The Logic of the Gaze*, perspective thought followed the logic of the fixed *gaze* rather than the unstable and shimmering *glance*, thus yielding a visuality that was reduced to a settled and single point of view and in this sense the fabrication of linear perspective was, and is, anti-immersive and anti-noise in disposition. Romanyshyn supports Bryson's contention.[205] Clearly this reduction of our actual wobbly vision[206] into one absolute point of view can only be achieved by negating the beholders' peripheral visual attention. Only by establishing the fiction of the viewer's partial absence and lack of glance can enthralment by fixed perspectivism be secured.[207] The perspectivist viewer is thus excluded from immersive participation in the art, held at bay as it were, and excluded in the interests of objectivity through the methods of exclusion and voyeurism. Correspondingly, the noisy world as seen by this immobile and atemporal gaze becomes stagnant, reified, fixated, inert and deadened.

In the Renaissance ideal of linear *trompe l'oeil* perspective, infinity, mathematics, and theology met on a unified plane whose grandeur and rational perfection symbolises a faraway, mighty and incomprehensible God. This perspectivist symbolization is explained in Erwin Panofsky's *Perspective as Symbolic Form* as the ideal image of infinite distance. The theoreticians of the Renaissance went so far in pursuing this ideal as to expurgate Euclid, omitting from their translations his 18th theorem which was inconvenient to their theory of ocular *perspectiva artificials*. However, the theory of perspectiva artificialis not only shored up the Quattrocento religious ideal of a distant Godly infinity but, as a result, enhanced the detached spectator, who (like God) exists and observes distance now from afar by isolating and cutting ambient vision off at its edges and retracting it to a frame. Viewing through the Renaissance intentional window, the onlooker holds an exclusive singular viewpoint and hence space becomes geometrically isotropic and rectilinear. We now have a detached transcendental subject constructed by ignoring the optic characteristics of immersive noise space and by repressing peripheral attention to the encircling atmosphere.

But not only supposedly transcendental in its ideological origin, this rectilinear vision also represented a nascent scientific understanding of the world that motivated the dissecting of optical immersive space.[208] The fragmentation of the noisy immersive world is now underway as the geometric grid divides and subdivides sight and the world into smaller and smaller manageable portions.[209] Of course, the vast majority of media images (and most visual art) produced today still cleaves to this horizon-line based Quattrocento framing operation, as opposed to the immersive noise span where horizon and frame dissolution is desirable. The invention of photography, and the astounding rapidity with which it spread, is closely connected to the fact that perspective, and its specific corresponding intellectual configuration, had pervaded visual habit since the Renaissance.[210] Renaissance linear perspective however, it must be remembered, is only a convention which, as Panofsky argued, is a cultural attribute comprehensible only for a quite specific sense of space or perception of the world and definitely not an absolute perceptual truth.[211]

Though Christianity primarily shaped the ideology of the period, no solitary philosophy or ideology dominated the cerebral liveliness of the Renaissance. Interest in neo-Platonic theories, the occult, sorcery, and astrology were widespread even as the authoritatively endorsed subjugation of magic began during the Renaissance. At the same time, Renaissance proto-humanist scholars and critics proclaimed that their age had progressed beyond the brutality of the past and had found its inspiration, and its closest parallel, in the civilizations of ancient Greece and Rome. By the 15th century, intensive study of Greek as well as Latin classical history gave Quattrocento scholars a more sophisticated view of antiquity and eventually Renaissance ideology spread north of the Italian Alps to all the courts of Europe.

However, we must keep in mind that the high Renaissance style in art (created by primarily Leonardo da Vinci (1452-1519), Donato Bramante, Michelangelo Buonarroti, Raphaël Sanzio, and Tiziano Vecellio Titian) endured for only a brief period, from about 1495 to 1520. To periodize, we can say that the High Renaissance ended when Mannerism began (about 1525) but this is only partly accurate, for all movements in European art up to J. M. W. Turner,[212] Impressionism, and particularly Paul Cézanne (1819-1877) depended on Renaissance spatial ideas,[213] a hegemony maintained for centuries by the widespread institution of the Art Academy.

One rare aspect of the art of the High Renaissance which has noise characteristics similar to those we previously saw in the Apse of Lascaux, is its search for a general, unified effect of *sfumato* composition, a smoky technique used for decreasing the separating dramatic force and physical presence of isolated figures in a work of art through immersing them in a fumey, semi-imperturbable equilibrium. Sfumato is the subtle, smoothly imperceptible, gradation of dark colors which approaches a smoggy unity useful in the creation of psychological atmospheric effects evocative of the immersive display in the Apse of Lascaux. Through sfumato, complimentary contrasts (*contrapposto*) find a unity previously absent and it is this unity that lends Renaissance vision its most significant self-alternative to the soon hegemonic point-perspective. This is so as sfumato invites and promotes an expanded, diaphanous, dilated focus and necessitates a more expansive field-of-view (which as we have seen is consequential for experiencing immersion into noise). Thus an immersive (anti-perspectivist) characteristic of high Renaissance art was *sfumato unity*, particularly because it depended upon a balance achieved as a matter of intuition and hence was beyond the reach of rational knowledge or technical manoeuvres. With sfumato, we see the seeds of an immersive noise counter-tradition in opposition to the crisp, detached, geometricized optics of point-perspective. This oppositional optic practice of sfumato visualization, which brings receptive vision to a state of sympathetic languor, Leonardo da Vinci taught to his students in his *Treatise on Painting*. There he encouraged languid attention to the ambiguous grubbiness of cracks and smudges on decrepit walls, which may suggest faces and forms to the viewer in order to aid artistic imaginative and visionary ability.

Thus sfumato offers another type of management of vision and an expenditure of the incognizant exploration of immersive noise excess.

However, a far more overriding artistic strategy was the pursuit of the ideal of "true" point-perspective which developed during the early-15th century (the early Quattrocento) in Florence. Filippo Brunelleschi (1377-1446)[214] is traditionally accorded the accolade of being perspective's pragmatic designer, as he created a sense of depth that integrated (by implication) the spectator outside the framed pictorial space. Brunelleschi was certainly the first to carry out a series of optical experiments that led to a mathematical theory of perspective. Unquestionably, Brunelleschi in 1425 contrived the first painting in "true perspective" when he insisted that his friends stand exactly where he himself had stood while painting the panel of the Baptistery of St. John in the Piazza del Duomo of Florence, and directed them to look upon the original scene he had painted. Then he held the painting up with its backside directly in front of the viewer's face. A tiny eye-hole was drilled through the middle of the panel. Gazing through the eye-hole, a viewer simply witnessed the original scene. But if a mirror was held up in front of the painting, the viewer now beheld the painting instead, and it was so accurately done in perspective that it was supposedly indistinguishable from the original.

Moreover, Brunelleschi analyzed human vision mathematically and by so doing discovered the suppositional central vanishing-point that the horizon-line passes through (which is also the line on which two-point perspective is defined by the oblique vanishing points). With this schematization begins the emanation of a perspectivist scheme for envisioning and depicting range that remains paradigmatic to this day. Such a *trompel'oeil*, linear perspective casts a system of single-point co-ordinates over the actual far-reaching manifold sphere, in the fabrication of an illusionism which deceives visible perception. This perspective tradition, according to Hal Foster, "was based upon the premise that the spectator's eye was singular, rather than as double as with normal binocular vision".[215] Hence, it represents vision through geometric perspective by projecting and holding holonetric 360° vision to a single, fixed-eye point and it is just these fixed rules of perspective that construct an *anti-immersivism* and creates and expresses anti-ambient divisions between the subject and the space.

Leon Battista Alberti is avowed as perspective's inaugural theoretical interpreter with the 1436 publication of his text *De Pictura* (On Painting). According to Edgerton, from that point on Western artists conceived of their subjects in terms of an imposed *spatial homogeneity* as determined by the horizon-line and the fixed gaze.[216] While it is possible to indicate the sources for perspective in Euclidean optics and geometry and in late-medieval/gothic versions of (and commentaries on) Arabic works on optics, the application of such theoretical material into the creation of a working system, and its transfer from the realm of physiology, philosophy and theology to that of painting's conceptual *window* is what constitutes Alberti's achievement. This achievement, however, also formulated a restricting boundary between the self and the richness of the noisy sensual world. Indeed this window, which was at first conceived as being open, eventually closed and became grilled with a grid[217] and it is just this closed and grilled optical perspective which, according to Romanyshyn, has become the cultural visual hegemony of the modern world.[218] As a result, John Berger summarizes Alberti's window in disparaging terms as a "safe let into a wall, a safe into which the visible has been deposited".[219]

Curiously, it is interesting to discover that Alberti designed grottoes modelled explicitly after ancient classical ones, which certainly cuts across any overly simplistic explanations of artistic ideals. However, Panofsky adequately demonstrated that Alberti and Brunelleschi, for the most part, tried to forget and obliterate the spherical allocentric visual noise field of the ancients, in favor of an angular-linear perspectivism by ignoring the bumbling of noise and glance.

So let's turn underground, shall we? When immersed within the sfumato ambience of the fecund and grisly baroque/neo-baroque Cimitero dei Cappuccini located beneath the chapel Santa Maria della Concezione in Rome (circa 1626 to 1870),[220] I found myself tipsy with a morose and peculiar noise vision of death: overwhelmed, engulfed, and supersaturated by an extremely dark lapidary style.[221] The *mises en scène* of this art-ossuary are arranged from innumerable human bones (the skeletons of over 4,000 monks) displayed by their survivors, the Alinari Brothers, in an unashamedly ornamental attack on the simplicity of death. The abominable, dank, subterranean syntax is so rich and evocative as to border on logorrhea. The style so purple as to spill over into ultraviolet.

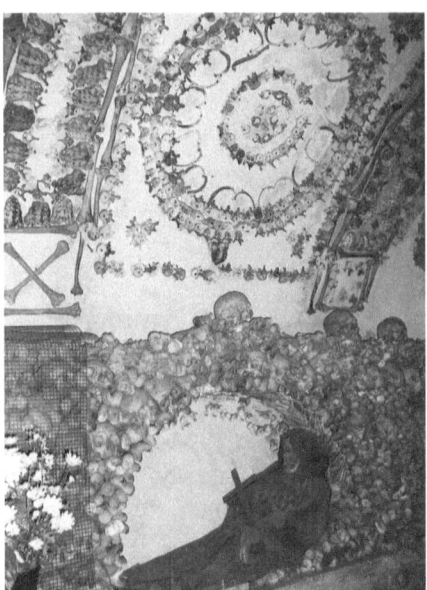

Figure 6 – Side ossuary in the Cimitero dei Cappuccini located beneath the chapel Santa Maria della Concezione (Rome)

This immersively opulent cove resonates with a beguiling virtuosic uneasiness which reaches far beyond its material circumference. And yet, at a time when our critical abilities have been systematically bastardized by media saturation and bludgeoned by decades of commerciality, the Capuchin monks' convoluted baroque visual style is to most merely disturbingly freakish. More to the pity, because the Capuchin crypt spoke to me with an art noise voice both compelling and concerted. Indeed the crypt is such a powerful immersive noise space that it thrashed me out of my scepticism about the limits of eerie fantasy as I felt about me a disembodied, skinless, and offensive silent stillness from where countless skeletons seemed to stir and quiver and seethe about abhorrently. The elaborate unity and horrid continuity effect was not one of melodramatic gloom, however, but of comic/tragic reconciliation formed under the aegis of a totalized noise excess.

The Cimitero dei Cappuccini's flower-arrangement-like nihilism is another example of the immersive noise dissolution of representation through negative hyper-promiscuity as discussed in relation to the Apse of Lascaux's occupied sfumato. Here, this promiscuity is further enhanced by emphasizing our human disintegration anxiety as the crypt collapses human identity into a construction made up of literally dead distinctions between selves and signs. This fabrication speaks to the fact that we are all but schematic information (genetic code) immersed within the millenarian field.

Equally, the crypt made me sense that the precarious glittering life of today's representations are made up simply of all the previous images they have succeeded in disintegrating and recomposing. The eye can scan and emotionally identify meaning in the perceptual field of the

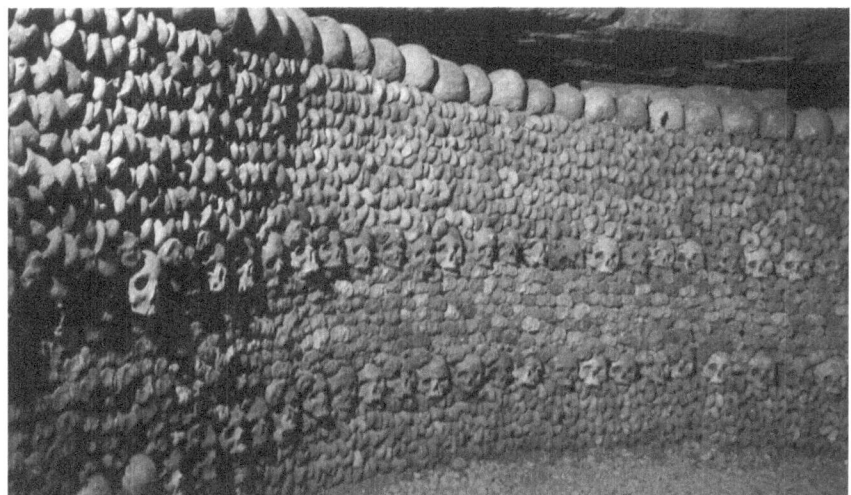

Figure 7 – Detail of a section of the Parisian Catacombs (Paris)

crypt only because its structure is the sfumato, concave side of our own personal ego-image. The crypt is in a sense, then, like the Apse, another noise representation of all representations. And, as such, it is an attempt to represent the unlimited immersive field of representation. Therefore, it urged on me the idea of an immersive noise space in which images no longer have any identity or distinctive place. Rather, here in the crypt's semi-chaos and ferment, lay great hidden forces. Forces of vital emotional release, where things and bodies are represented only from the madness and ecstasy which animate them. Here all are equally joined in the great flow of life and death as, in the depths of this compactness, blood, excrement, and doom join in noise obscurity. Bound now inescapably and tightly together, human forms—and the blank space that usually isolates them and surrounds their outline—interpenetrate each other in an immersive folly far more horrific than transcendental. Anything less strident, less terrifying, less crazy, less intoxicated, less contaminating to our perspectivist gaze would not be able to de/re-compose it, as it must be if we are to achieve the vacuole basis of noise cognition.

I followed up this dire saturation by visiting the usual early Christian catacombs in Rome and I found them powerful but entirely too barren to say more to this noise study. However, after perilously flying back into Paris in fear for my life through what in France is called a *tempête* (wind storm), I immediately went to visit the Catacombs of Paris, which I found rich in associative noise material that addressed existential anxieties in Baroque fashion.

The Parisian Catacombs were begun in 1785 following the decision to excavate all of the city cemeteries, stemming back to Gallo-Roman times, and move the remains to one central location, an abandoned Montrouge quarry from which the stones that built Paris were dug. Between five and six million skeletons were thereupon transferred to this location and stacked systematically according to body part (seemingly endlessly) against the walls. The walk through the resulting macabre grandeur, with its repetitive super-coded/anti-coded rigor, is stunningly beautiful. A noise rhythm of great discord is established, necessarily entailing a process of intimidating sublime ideas which foresee us in expiration along with our own minuscule insignificance. This is immersion into noise incapable of maintaining racial or sexual differences and, as such, is the animated crumbling of the normal monuments to human differences we construct daily. The Catacombs' tottering bone-lined promenade works against the completing, reassuring, mystifying representations of ideology and, as such, it ponderously invites the ultimate integration of form through an immersively indiscriminate account of human existence, complete with resultant long, painful, beautiful washes of commingled feelings and realizations of our ultimate maggot destination.

On returning to the light of day so as to inspect characteristics of baroque metaphysical noise space (clearly the Baroque is an ebullient Catholic art practice rather than a restraining Protestant one), I observed that the Baroque is saturated with the power of deception, with make-believe and *trompe l'oeil* effects. Indeed, it is teeming with the gilded complexities of ethereal flowers, throbbing stars, false arches, dripping fruit, twisting leaves, winding columns, floating virgins, spinning clouds, and resplendent angels. In Baroque noise vision, artistic expression was predisposed to try to break out of bodily place and architectural space by superseding forms of constraint by over-saturating the norm.

A good example of this noise tendency is the late-Baroque Catholic church in the tiny Mexican village of Santa Maria Tonantzintla in that it is an Indian version of Baroque flourish. Here, an all-over excessive decorative web dances around one in unrestrained profusion and forms seem to explode with pleasure as everywhere foliage glistens, leaves shine, angels hover, and carved fruit exude thick drops of dark honey. Such syncretistic excess is typical of the Late-Hispanic-Baroque, which is also called the *Churrigueresque*, after the Spanish architect José de Churriguera (1665-1725). This period has been called an exaggeration of the Baroque to such an extent that it concluded it.[222] Other fine Mexican examples of this tendency are the altar retablo in the former Jesuit seminary in Tepotzotlàn and the Rosary Chapel in the Church of Santo Domingo in Puebla. Here a decorative web extends beyond the retablo through the use of what is called *yeseria*, a type of interior plaster work which was used to cover vast areas in ornamental moulded relief. This technique evolved into what is called *argamasa* and was utilized widely in the Hispanic Querétaro style, which was basically Rococo by then. Both yeseria and argamasa provided a good base for the application of vivid color, a taste Mexican architects satisfied fully through their use of glazed ceramic tiles which were used to cover everything from building façades to entire domes and cupolas.[223]

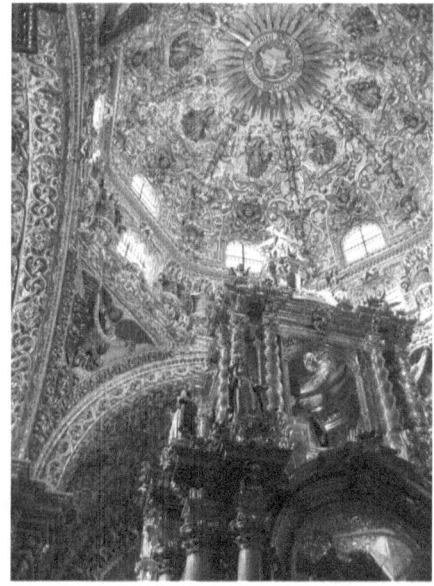

Figure 8 – *Rosario Chapel* in Santo Domingo Church (Puebla, Mexico)

As we have already seen, in the absence of previous framing restrictions, the Baroque is fraught with immersive noise challenges which may produce ecstasy in some or bafflement and mystification in others. Although it tolerated and accepted, even celebrated, the individual unique aspect of life, it transfigured all of its dispersed elements into a single unifying will. Hence, the noise world of the Baroque is a diverse one of contrasts while remaining essentially one unified (Roman Catholic) world. Most importantly, 17th century Baroque art flaunted a prismatic rejection of Protestant visual simplicity, thereby undermining the, by then, conventional clarity of perspectivism in that perspectivalism, as previously explained, sequesters the subject from the environment by constructing the subject as supreme and the space of vision as detached.

Central to our noise concerns is the baroque niche's format which links us back up to the grotto and all that it represents in terms of noise contemplation. Baroque niche-space continues the immersive noise impulse and the immersive disposition towards filling and overflowing peripheral ambient vision with noise excess which implicitly returns perspectivalism to its legitimate province as a narrow and contingent intellectual convention.

What interests in the High Baroque is just this scopic all-over tension, as it emanates a forced, but fine spun grace in a vigorous noise continuum. The High Baroque's multiple, disjunctive strands of meaning are presented to the spectator simultaneously and it is for the viewers' swift, sophisticated mind to create a single tissue of meaning from it, even as the body is engaged in movement and in opposition to the disinterested static gaze of the dominant but petrified Renaissance beholder. As we have seen, Renaissance space tended to be torpid and planar, its harmoniousness achieved through a cumulative addition of clearly defined cognate elements, while Baroque totality is achieved (at the expense of clearly defined elements) through the subordination of individual elements into an activated whole. In the Baroque niche, it is the inter-interpreting spectator, not solely the artist, who may be regarded as the agent who affects meaning-laden synthesis on the rhythmically continuous but diverse component parts inextricably interwoven there.

Also among the prevailing characteristics of the Baroque niche relevant to the immersive noise intention are its feelings for a stirring dynamism and a wide gesticulation. Emblematic of that Baroque ideal is the sensation of theater (involving the motion of the spectator) which is evoked.

This is the Baroque way to reshape bucolic experience by increasing seductive opulence while at the same time seeking fine and intimate sensations of diversity in the moving sensuousness of its unified means. With it, an aspect of the Baroque achieved a partial overthrow of the dominant Renaissance scopic order and the immersive is elevated to a position of momentary superiority. It is precisely this pernicious dereification of visual stasis that prepares the way for, and celebrates, the implications of the phantasmagorical aspects of immersion into noise, as in the Baroque period there existed a characteristic demand for phantasmagorical illusions which required that an interior space, in churches principally, consummated a remarkably effulgent sense of Gesamtkunstwerk totality by blending the elements of architecture, sculpture, painting, ornament and lighting together. The aim of this Gesamtkunstwerk fusing was to orchestrate an otherworldly consciousness, which was to enhance the beholder's faith through the creation of an effect of unitary intensity in opposition to a continuation of consciousness' normal segmenting function. Consequently, the phantasmagorically affected Baroque atmosphere of indeterminate apparition was considerably concerned with a seductive noise style which functions best as whole Gestalt composites which convey a sense of awesome grandeur, spatial complexity, and an interest in excess.

Though primarily Latin in sensibility (in service of the Roman Church's parsimonious and authoritative inducements), there are fine examples of the Baroque taste for unrestrained, phantasmagorically inclined ecclesiastical Gesamtkunstwerks in Germanic countries as well, including the Abbey Church at Melk, Austria by Jakob Prandtauer (1660-1726) and the Benedictine Abbey Church at Zwiefalten, Germany by Johann Michael Fischer (1691-1766). Secular architecture, too, aspired to the ideal of attaining Gesamtkunstwerk union between all of its parts and the building's surroundings as in the case of Gartenpalast in der Rossau (1711) which today serves us as the Museum of Modern Art in Vienna. The patron of the building, the prince of Liechtenstein, planned out both the building and the gardens with unified Gesamtkunstwerk intent. However the juiciest noise spaces are the sacred ones, as in the aesthetic overture of the Baroque, religious doctrine became the *modus operandi* of inducement. Hence noise affect was permitted to function through constructed illusions and through the enlargement of artificial inclination.

Previously, Renaissance church architecture had been principally sympathetic to an orderly architectonic system based upon the aesthetics of gracefulness. Beauty consisted in a prudent adherence to mathematical proportions in all sections of the building. In comparison, the Baroque church as a medium sought to establish a heightened perceptual instrumentality between the visualizing subject and that of an entangling *magnum opus*. It is this union of apportionment with visually stimulating interchanges that make up the sturdy but fine corpulence of the Baroque's exorbitant proposition. In contrast to the lucid, linear, fixed, sealed framework of the Renaissance perspective and world-view, in the

Cinqueccento Baroque we observe a complication of spatial levels so ambiguously interrelated and so multiplied as to leave no one single immutable plane of reference the spectator can grasp.

Philosophically, this Counter-Reformational submission to vertiginous noise experiences of rapture are indicative of the Baroque propensity for self-consciously eschewing the model of intellectual clarity in favor of a language of multiple ambiguities and shifting excess. We must recall that the *Reformation* was the reform of the Roman Church in the early 16th century which came from those who protested against its excesses. The various Protestant churches set up as a result of this reform ideal profoundly influenced the nature and scope of art where they flourished, and this impacted heavily upon the employment of artists. In reaction against the opulent (and hence expensive) excessive aura of the Latin (Mediterranean tinged) Church, Protestants favored strict simplicity. Hence church commissions declined. One can see how this would (or could) break up the artistic employment structure which was capable of producing immersive installations. Still, as immersive noise art in the North shrivelled, the Roman Church retaliated with the *Counter-Reformation* agenda, a vigorous counter-offensive running from about 1560 through 1648 which offered the public a new, even more, energetically excessive program of immersive abundance. With it, church construction and embellishment boomed, offering artists in Catholic countries a wealth of work. This counter-offensive initiated a revival of confidence in immersive noise experience which the Catholic Church, as it was now called, lavishly funded. Thus artists in Catholic countries worked, albeit attendant upon the narrow ideological objectives that allowed the art to exist. Happily, these objectives generally display a wider enchantment with nebulous noise propositions which it ostensibly attempts to delineate by making the actual physical medium almost nowhere admit to being only what it is, preferring to simulate other media such as tapestry and/or bas-relief sculpture. Thus, we are presented here with an illusionistic noise experience which shifts itself in a softly focused, multiple and perforated manner.

Accordingly, the unrestrictedness of the Baroque visual/intellectual situation goes beyond ideology towards a noise multivalence by way of a smoothed disjunction that supplies a unity of vision and fills the air with an attitude where space altogether ceases to be conceived as a void and becomes nearly palpable in its fused and responsive ether. Thus Baroque spatial composition results in creating not a clear, unproblematic, ideological art, but rather a dazzling and disorienting deftness by blending a surplus of images and forms into a majestic noise art. This Baroque dexterity inevitably weds suavity to grandeur through an implied sense of manifest splendor where elaborate conflicting contrapposta appears poised in equilibrium. Hence it provides a fluency and fullness to space which, when conceived of skillfully, becomes lyric and vibrant. As such, it creates a sensuous impression (though languorous) through an implied transition from analytic to synthetic comprehension of pictorial form which fundamentally marks the mentality of the Cinqueccento Baroque atmosphere: a détente mood to bind and unify forces. In this synthetic noise sense, then, the Baroque's rhizomatic visual injunction prepares art for the re-emergence of the immersive noise formation, in that it weans art away from the fiction of a "true" perspectivist visionality and reveals instead the possibilities which open up for inventing new scopic arrangements (and rediscovering lost ones).

Such a reduction in perspectivist constructs in favor of synthetic noise ideals comes together most vividly and succinctly in Baroque manifestation in the *bel composto* (beautiful assemblage) niches of Gian Lorenzo Bernini (1598-1680) which were constructed in the various baroque chapels he created in Rome in the late-1600s. Giovanni Careri's book, *Bernini Flights of Love, the Art of Devotion*, beautifully articulates what I have been feeling internally when struggling to define precisely what it is I mean by the instinct for noise in art. Careri analyses the synthesizing character of three Baroque chapels which Bernini assembled, often in terms of the montage film technique pioneered by Sergei Eisenstein (1898-1948) in the early part of the 20th century. My main concern, however, is with his analysis of the Fonseca Chapel within San Lorenzo, Lucina in Rome and the immersive noise which Bernini articulated through the, at that time new, *bel composto* mixed-media technique. If fact, my interest is precisely located in the upper third portion of this composition, the segment concerning the relationship between the angelic bodies and their position in and relation to space, weight, and light. The way in which the angels first depicted in the top third in painting are extended beyond the limits of the painting into the garland of putti—which floats in the vault of the dome, linking the painted sky with the sculpted one, up into the central oculus, which reveals the actual sky—presents an interesting progression in the expansion of the frame and invites issues of immersive noise to come forth.

In order to understand how noise ecstasy is represented in the rippling undulation of the *bel composto*, we must see how painting, sculpture and architecture are "linked together in a fluid ensemble designed to create the experience of an overall expanding frame of reference" based on inter-relationships assembled between "miscellaneous hierarchical arborescent perceptions"[224] in correspondence with their position in the sum synthetic-noise-total, as well as the contingent location of the spectator. These correlations, which guide us "through the composto by the montage of the arts",[225] were explicitly announced by Bernini himself on his pronouncing that things in the Fonseca Chapel do not appear only as they are, but also in relation to what is near them—a relationship that changes their appearance.

To best understand the noise issues at work here, we need to look again at the metaphysical underpinnings driving the artistic expression, specifically, the metaphysical ideology behind conceptions and representations of angels. Far from being the quibblings of cloistered theologians which bore no relationship to life, these concerns involved the very bedrock of the Church's theological, cosmological, and philosophical structure and teachings, as the figure of the angel was a primary representation of the human's position concerning relationships between space and matter in expansive/immersive terms. As such, a consideration of the angel's efficacy in an idealized condition sets down an influential model for human potentiality.

The term *angel* is derived from the Greek word *angelos* which means *courier*. In that the messages delivered are airborne and move, angels fly and are winged. When we say that an angel is in a place, we mean that (s)he has applied *virtus* (an inherent power and potential) to that place. *Virtus* means both the potentiality and the capability to *generate special effects*. In the Koran, every angel is the key to a different endless ocean of knowledge which has no beginning and no end. Yet the exact composition and material quality of angels' bodies and how they relate to the space of the world and the celestial space of heaven, had remained an urgent concern and of great debate in the Christian Medieval Ages between Augustinian Franciscans, followers of Saint Augustine (AD 354-

430) and the Aristotelian Dominicans, followers of Thomas Aquinas (the Italian theologian who largely adopted the neo-Platonic ranking of angels believed to have been written by Dionysius the Areopagite). Augustinian Franciscan Saint Bonaventure (1221-1274) argued against the Aristotelian Dominicans that no *pure* form existed in creation and that the angels were composed of *hylomorphic matter,* matter bound up with the principles of otherness and mutability. Thus, since matter contains the principles of mutability, angels posses a quasi-material body (what to me seems to be an evocation of the fat-light of white noise). This fat white noise angelic body cannot be circumscribed by place, however, because place is a quality whereas quantity is endowed with position, unlike the quantum nature of light. Following the Bonaventurian model, the angels portrayed in Bernini's Fonseca Chapel imply in fat-light fashion that their bodies are constructed from a hyper-sensitive semi-materiality, fabricated of hylomorphic semi-transparent matter.

Primary to the Bonaventurian angelology of the 17th century were the above distinctions, distinctions which were meant to also counter the Protestant critique of the Holy Eucharist. Hence, angels became the paradigmatic representation of immaterial hylomorphic substance. The main point of the Bonaventurian/Augustinian angelology—and what is at the hub of this reflection—is the idealization of the semi-transparent hylomorphic flesh of the angel and its location and existence in space as being material yet virtually interfaceable with immaterial spatiality, in other words its possession of *white noise viractuality*.[226]

The permeable quality of an angel's hylomorphic white noise body is represented in Bernini's Baroque Fonseca Chapel's *bel composto* by the child-like winsome figures' semi-transparent relationship to their painted background, which conceptually spills over into the painting's sculpted frame and into the architecturally domed space from which the angelic figures emerge and return towards a light which constantly re-defines them with every fluctuation in its intensity. From the painted portion of the *composto* (by Giacinto Gimignani (1611-1681)) the garland of hylomorphic angels is transposed into relief sculptures that overflow the frame and expand the painting as if they were released from it into the space of the architecture en route to, and from, the circular domed light source which dominates the composition and both physically clarifies and luminescently dissolves the depicted hylomorphic forms.

This conveyance of semi-transparent hylomorphic entering and exiting by preternatural means is the role the angels play in the Annunciation narrative, which depicts and explains the forecoming pregnancy of the virgin Mary below. The inference of these hylomorphic childish forms emerging from and returning to a central radiant hole stresses the narrative of female sexuality and reproduction (just as in the Pagan grotto) and hence again brings forth immersive noise issues.

On entering the church on a bright Italian day and approaching the dim Fonseca Chapel, the dilation of the pupil of the eye in reaction to the abundance, then semi-absence, and then increased presence of light within the sombre enclave (in parallel with the circular overhead oculus (reminiscent of the Pantheon) from which the heads and faces of angels peek) harmonizes spectacularly with the dilation process of the female sexual aperture that precedes and precludes sex and birth. In the Fonseca Chapel's *bel composto*, childish figures emerge out of the oculus and paintings, and float in a gravity-free environment in which their tiny nude bodies break free from, or are consumed by, light. The individual hylomorphic bodies that construct the spiral of cherubs which leads to and from the oculus in the dome grow progressively diminutive in relationship to their location near the oculus, creating the impression that their bodies are penetrating the stucco material creating the dome, as if the architecture had no more physical density than a hylomorphic fat-cloud. The hylomorphic materialization of the cherubs reaches maximum transport the closer it is situated near the light emitting perforation, and it is precisely here in the Fonseca Chapel's *bel composto* where white noise viractuality beckons the flesh to go outside of itself breath-like and for spirit to abandon the sheath of rational flesh in ecstasy. Here the membrane of ecstatic and swooning flesh is submitted to luminous pressures from within and infusions from without as it resists not the sacred/orgasmic passion of expanding and then re-assembling the self in a continual, mobile, dilation-immersion into noise into *ex-stasis* (which literally means *gone outside itself as standing still*) typical of the symbolism of the cave/grotto. As Careri asserts, "in contemplation, the composto tends to *go outside itself,* in its own ecstasy becoming the vehicle of an experience that goes beyond all images".[227]

This *ecstasy of going outside of self* (with its breath-like mercurial countenance) defines one important art noise ideal. Aesthetic immersive noise spaces (as we have seen particularly in the Apse, the grotto, and now the niche) give license to this particular dynamic dialectic of going outside of self in a way the perspectivist tradition seems incapable of doing, as the immersant is better arranged in symbolic space to voyage the unveiling circuits which are employed in the encouragement of ecstasy than when facing the pictorial. For when positioned within an immersive/expansive noise field, the immersant's ambient vision is already being drawn peripherally outside itself and outside its commonly restricted (framed) edges.

Hence the Fonseca Chapel's encouragement of floating transference is the tension of a representation outstripping itself and, as such, it produces a noisy rapturous effect as it continually over-leaps the scope of image and concept and identity and space where one capacity turns against another in supernatural contradictory fashion, reminiscent of the ecstatic writings of Saint Theresa of Avila (1515-1582) and her descriptions of an engaging mystical union with Christ. While Saint Theresa describes this mystical union as a wave-like experience that cannot be related outside of the terms of incomplete and opposing images, she nevertheless stresses the experience as being one of a dynamic, intense and convulsive nature.[228] These attributes, as codified in the Carmelite spiritual tradition, became another semi-transparent model on which Bernini based his teeming composition's rippling and pulsating operational mechanisms, including his Cornaro Chapel *bel composto* sculptural niche in San Maria della Vittoria which embraces his famous *Ecstasy of St. Theresa* (1652): what, frankly, appears to be a woman, tucked into her grotto, in the throes of orgasm.

Consequently, within Bernini's Cornaro Chapel niche, the boundaries separating the ecstatic noise realm from the unecstatic realm are transgressed convulsively by St. Theresa's semi-transparent relationship to wave-like space and omnijective-like corporeality. In the Cornaro Chapel niche, the writhing St. Theresa represents human (quasi-sexual) ecstatic potential as opposed to our relatively inviolable demeanour. Here, too, by being able to pierce solid matter in analogy to the way veins web marble, the angelic cortege is privileged to escape the rules of containment that matter imposes upon the non-hylomorphic human body. In trespassing the mortal boundary, Bernini's angels are again unplatitudinously associated with a broad electromagnetic spectrum of fat white noise and hence stand in opposition to containment. This makes the hylomorphic angel the site of hyper-real hyper-planarity and the perfect model to analyze art in terms of noise and extensions of the subject into an implicit space reminiscent of what in magic is called entry into the *imaginable-world*. This imaginable-world is more subtle than the earthly world and yet denser than the angelic world (thus open to white noise-like interactions) and, hence, an art noise zone of connections and ecstatic flights of release, even as the body is clamped to the heavy, round, earth realm which is holding everything down with its transparent gravity.

Rococo Noise: Between Opulent Habitat and Non-Place

We shall now switch epochal emphasis from the time of the rich fullness and dynamic noise extravagance of the Counter-Reformationist Baroque period to that of the Rococo. Even more than with the Baroque, the Rococo spills its vast organized commotion over the viewer, assaulting the immersant with commands for his or her aesthetic involvement in its artificiality and elaborate unreality. Unfortunately, photography flattens what in the Rococo is a complex but light application of ornamental splendor by overemphasizing details that are only part of a whole fluid spatial strategy, which serves to create a sense of amazement—a sense which these spaces still exude today.

Assuredly, amazement is the initial impression. Nevertheless, turgid or ecstatic noise seem to be the other two ways of interpreting the density of extensively adorned rococo space, according to taste. What is factually agreed upon is that the term *Rococo* came into usage in the closing years of the 18th century, although it was not acknowledged officially by the Académie Française until as late as 1842 when the Académie Française's supplement to its dictionary defined it as ornament characteristic of the reign of Louis XV (1710-1774) and the beginning of that of Louis XVI. The term was originally a derogatory one, most likely derived from the term *Rocaille*, which, as I described in the portion pertaining to the grotto, was a stylistic extenuation of craggy design used to fabricate interior grottoes and which tended to spread throughout the entire space. Briefly to recap: the florid sprawl that appears in the classical Greek nymph's sacred grove developed into the Italian grotto, which developed into the Renaissance interior grotto, which then passed into the fully all-over immersive noise attributes of the rocaille style. What interests me especially about the flamboyant vine-like sprawl of the Rocaille/Rococo interior, is how it suggests the viney rhizomatic shrubbery of the grotto entangled with our own bodies' thalamo-cortical system (the reticular activating system and the thalamus which links with the cortex, the basal ganglia, the hypothalamus, the hippocampus), as linear complexity is the motivating principle leading to the immersive noise effect in the Rococo room.

In majestic Rococo rooms, exuberant linear ostentation pushes visual logic mercilessly to its limits via opulence in an effectual fusion of intensity with grace. That is their *raison d'être* and their oppositional stance in relation to the coherent monocular geometricalization of the Renaissance point-perspective precedent. In fact, the sumptuousness of Rococo sensorial space is, one could say, the first mature historical paean of ecstatic noise, as in its architectural interior decoration the Rococo exhibits a fondness for the expanding all-over. With the ubiquitous asymmetrical curves of its surface structure, it wraps itself around and transforms an interior into a noise confection of delicately colored vines and ribbons, reminiscent of espousal pastry, and, only slightly removed, of the sacred grove of the nymph.

In 1683, the Premier Architecte du Roi (chief architect to the King of France) was Jules Hardouin Mansart (1646-1708), the architect who created the Hall of Mirrors in Versailles. It was in his studio and under his name that the Rococo style began to emerge, most keenly through specific motifs and innovations of his designers Pierre Lassurance I (1655-1724) and Pierre Le Pautre (1660-1744). By the beginning of the 18th century, the intermixture of the Baroque with elements of Classicism (which had characterized the royal art of Louis XIV) gave way to the lighter, more gleeful art of the Rococo, a spatial art which demonstrated a heightened immersive opulence. The spatial conception of Rococo architecture under Mansart began to show fundamental changes from the assured building of the Baroque. Where Baroque walls, piers and columns had been massive and forceful and where the space was tenaciously focused while being multifarious, Rococo spatial ideals, envisioned space as delicately unified while being diffused with an abundance of light. Stylistically, while the Rococo protracted the complexities of Baroque surface arrangement, it treated it as homogeneous ornamentation, justified only insofar as it charmed the eye. This stylistic shift away from the legacy of the rich Italian Baroque interiors of Pietro da Cortona (as first introduced to Paris by Giovanni Francesco Romanelli in his work of 1644 for Cardinal Jules Mazarin) began to introduce a lighter style (subsequently called Rococo) into the remodelling schemes of some of the rooms at Versailles. In the last year of Louis XIV's reign, this Rococo style was adapted in many interiors which involved Le Pautre's leading participation, including Parisian *hôtels particuliers* (private mansions) such as the Hôtel de Pontchartrain interiors built in 1703 and the chapel at Versailles, finished in 1710.

With Louis XIV's death in 1715, the characteristic phase of Rococo called *Régence* emerged. The all-over rippling watery feel introduced by Pierre Lassurance I and Pierre Le Pautre under Jules Hardouin Mansart was coupled with a new flashy plasticity consisting of curvaceous forms, mirrors, and oval hemicycles mixed with fluttering ribbons and/or acanthus leaves scrolling outward around a chamber. This trait is seen, for example, at the Hôtel d'Assy of 1719 which was constructed under the new Premier Architecte du Roi, Gilles-Marie Oppenord (1672-1742). Another bold régence style Rococo interior, known for its playful lightness, is the Hôtel de La Noiseilliére (now the Banque (Bank) of France) constructed by François-Antoine Vassé (1681-1736) who became the chief designer on Le Pautre's death in 1716. Further developments in the style were made by Condé architect Jean Aubert (1719-1785) who remodelled the Grand Château at Chantilly. In the Chambre de Monsieur le Prince and the Salon de Musique, Aubert extended gold filigree across the expanse of white panelling, along with a spidery scrollwork that spumed out onto the ceiling above the cornice at the corners and midpoints of the walls, creating a decisively noise effect.

Following the régence style of Rococo is what is called the *Rocaille* (the most extravagant expression of the Rococo) even as this term predates the emergence of the Rococo and suggested to it its name. As stated, Rocaille had originally referred to the shell-work employed in garden grottoes, but as of 1736 the term began to be used differently to designate a High Rococo, total, over-all design which included complimentary furniture and porcelain. La Pautre's basic concepts had been little embellished by Jean Aubert until Juste-Aurèle Meissonnier (1695-1750) and Nicolas Pineau (1684-1754) goosed the design concept into an even finer feathery network of gilt filigree that became ever more total while remaining light in spirit. Meissonnier, who was trained as a goldsmith, found his greatest fascination and inspiration in the asymmetrical. Although he produced only a small amount of work (his mark in fact has been found only on one piece: a gold and lapis lazuli snuff-box (1728)), his influence was widespread due to the favor of Louis XV and the posthumous publication of his designs in 1751 as the *Oeuvre de Juste-Aurèle Meissonnier*. In actuality, he had been responsible for only a handful of interiors: one for Léon de Brethous in Bayonne (1733), the apartment of Baronne de Bézenval in Paris (circa 1736) and a cabinet for Count Franciszek Bielinski in Dresden (1734).

Nicolas Pineau, on the other hand, was responsible for the more high-profile interiors of the Hôtel de Rouillé (1732), Hôtel de Villars (1733), Hôtel de Roquelaure (1733), and Hôtel Mazarin (1736) in Paris. Following the publication of Rocaille theoretical treatises, ornamental pattern sheets and suites of engravings by the Augsburgian publishers Martin Engelbrecht (1684-1756) and Johann Georg Hertel (1700-1776), the Rocaille ideal rapidly diffused throughout Europe and developed into what amounted to an international style. But by 1740 the Rocaille Rococo had reached its apogée in France and was universally accepted and hence underwent no further development. The most important works undertaken in this late phase are interiors by Germain Boffrand (1667-1754), who executed them late in his life. Notable was his creation of the Salon Ovale for the Princess at the Hôtel de Soubise (1739) where he blended a rhythmic succession of arched mirrors with extending tentacles of filigree with a series of painted panels by Charles-Joseph Natoire (1700-1777) placed in the spandrels and undulating coving, all spiralling around a central rosette.

Beginning around 1725, the Rococo held sway in Germany with even more strongly inscribed noise peculiarities than existed in France. This Bavarian Rococo is fantastic and more varied in form than its French inspiration, while being less courtly. The first Germanic architects of this style were Johann Balthazar Neumann (1678-1753) (particularly good examples are his amazing chapel at the Prince's Residenz at Würzburg (1741) and his church in Vierzehnhiligen) and François de Cuvilliés (1678-1768) (his odium in the Residenz is outstanding as is his pavilion at Amalienburg) in Bavaria and Georg Wenzel von Knobelsdorif (1607-1753) and Carl von Gontard (1738-1802) in Berlin. In Bavaria and Aus-

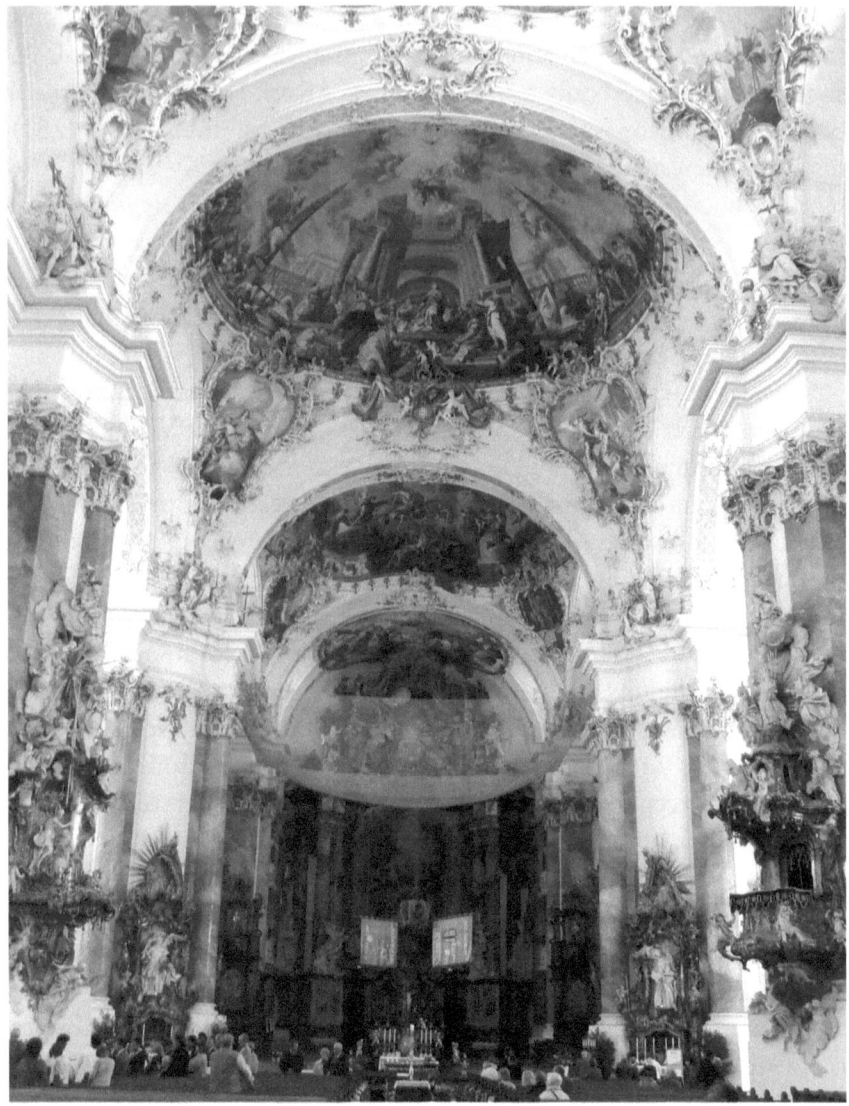

Figure 9 – Rococo interior of the Ottobeuren Abbey (Bavaria)

tria, the Rococo survived until the end of the century, while in France it had given way to the new austerity of Neo-Classicism by the 1770s. Very anti-noise.

A significant theme in Germanic Rococo counter-reformational excess is certainly an extenuation drawn from the Baroque expanse and its immense trembling flair for the plethoric but now taken to an even finer sugary spun *artiificialia*. For example, in the Church of the Assumption in the unassuming village of Rohr in Bavaria, *trompe l'oeil* theatrical curtains divide to disclose an absorbed Virgin Mary floating aloft, wanting any discernible means of support, reminiscent of the thinly veiled grotto theme of the Fonseca and Cornaro chapel niches. The Bavarian Rococo offers many such voluptuous grotto scenarios embedded within an architecture where structure melts into a bewildering myriad of curves and illusionistic spaces which give us pause, not just in the elaboration of viractual action, but in the congruent blending, flowing and folding of spatial pleats which forms an intrinsic part of a realm of experience recognisable as immersive noise.

Another marvellous example of Bavarian Rococo is the St. Johann Nepomuk Church in Munich (1746) which is better known as Asamkirche (Asam Church) after its architect Egid Quirin Asam (1692-1750). With his brother, the painter and architect Cosmas Damian Asam (1686-1739), he created a masterpiece of sumptuous noise at 32 Sendlinger Straße.

On entering the vestibule of the church, I encountered a consummate example of Bavarian excess. In this hybrid space, painting, sculpture and architecture work together in fabricating something between a prodigal odium, a playhouse, and a quixotic grotto. I was overwhelmed by a devastating folly of munificence and the giddy embellishment of silver and gold extravagance.

The Asam brothers also worked on the interior of the lush Einsiedeln Abbey south-east of Zurich. Another fine example of Bavarian rococo interiors can be seen at the Schloss Nymphenburg, near Munich, which was designed by Johann Baptist Zimmermann (1680-1758).

Further pertinent insights into immersive visual noise under the Bavarian Rococo are best characterized by the Wieskirche (1746-1754) designed by Dominikus Zimmermann (1685-1766) with its elaborate (possibly ostentatious) spectacular interplay of painted and sculpted forms. This church can stand representative of a dozen such 18th century whipped-cream ornamental eruptions in obscure villages, deep within the Bavarian, Swabian, and Franconian countryside: villages such as Ottobeuren, Weingarten, Osterhofen, Wallfahrsirche, Neresheim, Bobingen and Vierzehnheiligen.

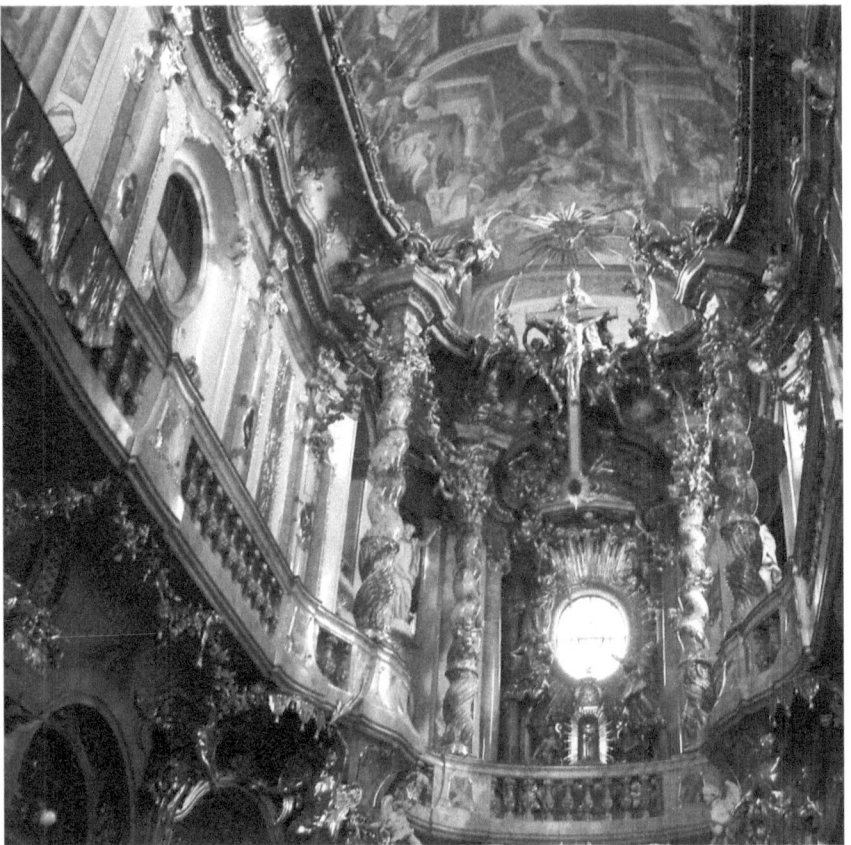

Figure 10 – Interior view of Egid Quirin Asam's *Asamkirche* (Munich)

The Neo-Rococo Noise of the Dream King

Typical of 19th century Neo-Rococo noise vision is the belief that all aspects of a comprehensive architectural scheme—from its landscape setting and the building itself, to the interior decorations, right down to the utensils—should be orchestrated as a seamless and homogeneous whole under the direction of one overriding design. This is the most enduring legacy of Rocaille style as its all-over objective became preserved and further elaborated in the Neo-Rococo. As we will see, it is a noise vision ideal that entwines its way through Fin-de-Siècle (1880-1899) architectural theory into one of the 20th century's driving art objectives. This complete integration within a constructed space of the broadest concepts on down to the smallest details (each reinforcing the other) is what is referred to as the Gesamtkunstwerkkonzept (concept of the total-artwork), a term adapted from Wagnerian operatic theory. The philosophical understanding of the canon of the Gesamtkunstwerk was the proclivity towards an integration of all related elements into a single aesthetic statement, resulting in a self-contained immersive world of *total design*.

King Ludwig II of Bavaria was born crown Prince on the morning of August 25th, 1845, eldest son of King Maximillian II (1811-1864). It is significant that he was born, and spent some of his early years, in the previously mentioned Nymphenburg Schloss replete with its rococo rooms, grottoes and frescoed scenes from Ovid's *Metamorphoses*. In 1857, at age 12, Prince Ludwig heard of *Lohengrin*, the operatic production by Richard Wagner (1813-1883) which was in production in Munich. That Christmas, Prince Ludwig received a copy of Wagner's 1851 text *Opera and Drama* from one of his tutors and soon after became captivated by all of the composer's published theories, including "Das Kunstwerk der Zukunft" (The Art of the Future) in which Wagner theorized the Gesamtkunstwerk.[229] This Gesamtkunstwerk ideal, in one way or another, affected the aesthetics of every one of Wagner's works from *The Valkyrie* on; including *Siegfried, Twilight of the Gods, Tristan and Isolde, The Mastersingers,* and *Parsifal*. It is this Gesamtkunstwerk desire for what Wagner saw as *total drama*[230] that was passed on to Prince Ludwig.

In 1861, Prince Ludwig saw his first production of Wagner's opera *Lohengrin* which made a profound impression on him (as it did on Wassily Kandinsky[231] (1866-1944)) instigating a long and intense admiration and eventual supportive role for the composer and his Gesamtkunstwerkkonzept ideals.[232] When Prince Ludwig was nineteen, his father, King Maximillian II, died unexpectedly, marking the beginning of the reign of King Ludwig the II of Bavaria, the *Dream King*, palace builder and generous patron to Richard Wagner.

After visiting his creations, it is safe to say, I believe, that King Ludwig desired a noisily decadent art for himself in ways of taste and state of mind. His aim of creating an inorganic world of excess and luxuriating in its rarefied artificiality was well articulated in 1884 with the publication of Joris-Karl Huysmans's (1848-1907) decadent novel *A Rebours* (Against Nature), a story of a reclusive art worshiper who yearns for new sensations and perverse pleasures within a transcendental artificial ideal. Decadent French theory, which is almost equivalent to Fin-de-Siècle

Symbolist theory, aspired to set art free from the materialistic preoccupations of industrial society.

Linderhof, one of King Ludwig's decadent fantasy palaces, was built in neo-Rococo style by Georg von Dollmann (1830-1895) to resemble the Petit Trianon of Versailles; Marie-Antoinette's (1755-1793) famous royal playground that was designed to resemble rural Austria (an impressive immersive work in itself) which included an adjacent Temple of Love. Linderhof is the only one of Ludwig's palaces that was actually finished. Of Linderhof, King Ludwig said in a letter: "Oh! it is essential to create such paradises, such poetical sanctuaries where one can forget for a while the dreadful age in which we live".[233] Located close to another of the King's castles, Neuschwanstein (designed by Eduard Riedel (1813-1885)), the King often retired to Linderhof to indulge in his decorated isolation. Linderhof owes a large part of its charged enchantment to the sublime natural beauty of its mountain setting and to its admirable prim French gardens. In the middle of its grounds, an embellished fountain emits a 30 meter high (about 100 foot) water-jet bathing a golden statue of Flore. The interior of Linderhof is a melée of neo-Rococo ostentation and mirrors (Bavarian Neo-Rococo is based on Bavarian Late-Rococo, an already plenteous noise style) and the glitter of gold is prevalent throughout. The King's Throne Room, modelled on an abstract Byzantine basilica, requires brief comment as King Ludwig oversaw every detail of its conception and execution. Its walls are arcaded on two levels and the ceiling suggests the immersive umbrella of a star studded cerulean stratosphere, with indigo, porphyry and gold as its predominant colors. Yet the most dazzling of the rooms are the Mirror Room and the King's bedroom (which were based on designs by Eugen Drollinger (1858-1930)).

However, it is another extraneous space—close by his lavish polyglot palace at Linderhof—which holds the most noisily significant (and cheeky) of King Ludwig's decadent dream realizations: the flamboyant Venus Grotto (a reference which brings us back to sacred nymphaea).

The 9.9 meter high (33 foot) Venus Grotto was designed by Fidelis Schabet (1813-1889) and fabricated in 1877 of garnished grout.[234] It was equipped with artificial arc-lighting, an ersatz rainbow, a wave machine and central heating, all set in harmonious action to recreate the phenomenon described in Wagner's first act of *Tannhäuser*.[235] The Venus Grotto was first intended to be built at Neuschwanstein, but due to lack of a suit-

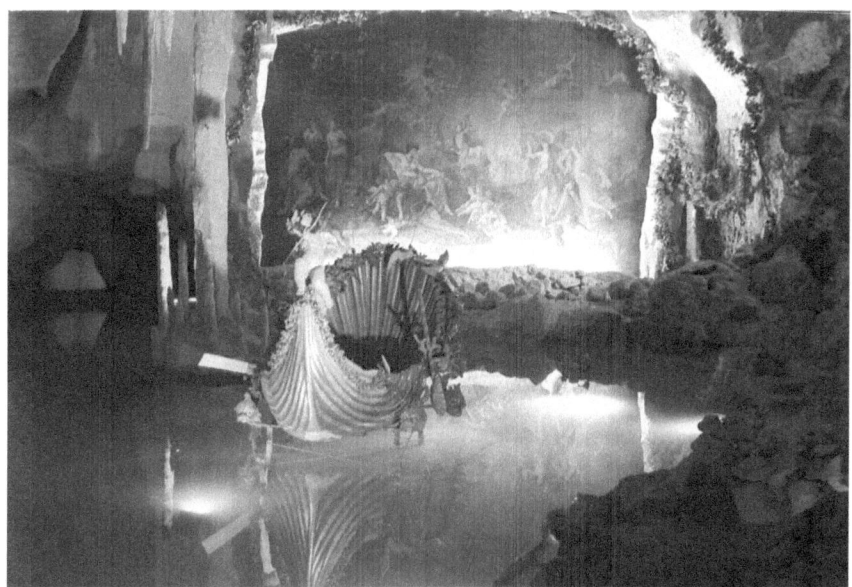

Figure 11 – Fidelis Schabet's decadent *Venus Grotto*, 1877 (Linderhof)

able site, it was moved to Linderhof by a December 15th, 1875, Royal decree and the work was carried out in 1876 and 1877. Dr. Michael Petzet, writing in Wilfrid Blunt's book *The Dream King: Ludwig II of Bavaria*, describes the grotto's space as one which allows the visitor an encircled mirage where "stage and auditorium are blended into one", creating a *total theater* as it "did not separate the onlooker from the stage".[236]

The Venus Grotto is furnished lavishly with fake stalactites, giving the impression that one has entered a Lascaux-like sacred noise space. According to Naomi Miller, this artificial grotto, compared to all others, "most nearly simulates the experience of exploring large cavernous spaces"[237] even as garlands of roses are strung throughout its 9.9 meter (33 foot) high cupola expanse (which extends hundreds of meters/feet inward). The grotto also contained a cascade and a fully functional artificial moon, and could be illuminated by electric lights colored to suit the mood of the King. The explicit models for the Venus Grotto were the Blue Grotto at Capri (Richard Hornig, the King's equerry, was sent twice to Capri to check the precise shade of blue) and King Maximillian's tiny grotto at Hohenschwangau, in which Ludwig had played as a prince.

In the Venus Grotto, five distinctive lighting effects could be made to play for ten minutes in turn by automated means, concluding with the appearance of a spectral rainbow just over the *Tannhäuser* set painting. It was very modern in that it was the first electrically illuminated installation in Bavaria. Of it King Ludwig said, "I don't want to know how it works. I just want to see the effects".[238]

King Ludwig had installed in his den (a study filled with paintings illustrating the erotic aspects of the *Tannhäuser* fable) a clandestine inlet which discharged him into his cherished Grotto of Venus. The Grotto of Venus is entered by a sharply angled antechamber which leads to the principal chamber. The first entity that one notices is a diminutive lagoon (replete with painted water nymphs, dryads and flying harpies) fed by a pattering cascade. As mentioned, the lights could be controlled to change colors, for instance to the cerulean of Capri or crimson to evoke the Grotto of Venus in the Hörselberg grotto where Tannhäuser dallied with the Goddess of Love. Exit from the grotto is made by way of a prolonged serpentine, stalactite-filled corridor which leads to a dolmen-like shaft that swings unclosed.

By gliding in the enchanting flamboyant cockle-boat over the face of the lagoon, King Ludwig could site himself in the midst of the grotto's ambience and surround himself entirely on every side, even if he was only experiencing a presentation that incorporated the variance of colored lighting effects. On the lagoon, which could be ruffled by an artificial wave machine, the King kept two swans, symbols of eternal bliss and immortality, along with his enchanting cockle-boat in which he would be rowed by a lucky servant.

Wilfrid Blunt, in his book *The Dream King: Ludwig II of Bavaria*, reports that there was a staging of the first act of *Tannhäuser* in the grotto but that the sound of the waterfall rendered subliminal the singers voices in a thick din mixture of sound and by an acoustical space described as "freakish".[239] This noise concert, however, Blunt mentions, may be fictitious.

CHAPTER 4

Modern Nervous Noise Eyes

The Imprudent Immersive Noise of the Fin-de-Siècle

As we saw with the Neo-Rococo, the main point of psychic noise through decadent design is the concept of space as a unified fuzzy entity. This concept of molding space into a unified whole through all-over noise marks the beginning of the trend away from eclecticism and towards Art Nouveau Gesamtkunstwerk ideals.

Art Nouveau is the French name of an art movement that took its impulse from the nymphianesque blend of natural forms and women. This interest in sensual bio-structure was expressed in sinuous Gesamtkunstwerk fashion, touching everything from cutlery to lamps to furniture to walls to entire building façades to metro-stations. Architects and designers who contributed to the development of this style included Victor Horta, Henry Van de Velde, Antoni Gaudí, and Hector Guimard (discussed below).

Basically, Art Nouveau is a Northern European and North American style of art/architecture which spanned from the 1890s to about the Neo-Neo-Classicism of the First World War era. It was called *Stile Liberty, Jugendstil, Modernism, Nieuwe Kunst* or *Sezessionstil* respectively in Italy, Germany, Spain, the Netherlands and Austria, but in all cases, artists and architects wanted to expunge the differentiation between the major and minor arts in the creation of a *total art* in Gesamtkunstwerk fashion by centring them around life. Therefore architecture, which has an immediate immersive sway on human existence, was the prominent art to which every artistic propensity is thoroughly integrated.

For our concerns, it is especially interesting to note the importance of Art Nouveau's approach to conceptualizing Gesamtkunstwerk interior space. As mentioned, the basis of the Art Nouveau interior is a concern with nymphian effluvium-feminine forms, and with swirling, tendril-derived patterns that are applied throughout the space in a frivolous spirit. The forms from nature most popular with Art Nouveau designers were characterized by flowing curves of the sacred grove: grasses, lilies, vines, and the sensual curves of women. However, on occasion, other, more unusual natural forms, were also used, such as peacock feathers, butterflies, and insects, but at all times High Art Nouveau's foremost feature is an emphasis upon ornamental value distributed throughout the entire space.

An Art Nouveau interior is asymmetrical at root, as evident in the tiniest single line or in its approach to the total space, but its typical asymmetricality is always in service of a *total design*. So what is important to our concerns, is that Art Nouveau is a noise art concerned with every detail, as every object of or in an Art Nouveau space is ideally related to a *homogeneously noisy whole*.

The obscurantist mystification often sensed in circuitous Art Nouveau was part of a widespread cultural reaction against the new social divisions brought about by the power of the Industrial Revolution and towards the intractable powers of the nymph/fairies at flippant play in nature. Its sinuous space provides the immersant with the possibility for an ebbing of consciousness toward the incomprehensible, a vantage point from which to breakout of the Renaissance perspective position towards a more supple non-Euclidean noisy awareness. This heightening of perceptual sensitivity allows for and encourages a heightened consciousness of one's surroundings in general, as the churned-line is found on the floor and then picked up in the shapes of the furniture and on into the doors and door frames until it reaches the structural arches which support the ceiling and into the lighting fixtures. As a result, the entire space is swaying, bending, floating, arching, smoking, curling, throbbing, dripping, melting, aching, writhing.

Baron Victor Horta (1861-1947), a Belgian artist/architect and teacher at the Brussels Vrije Universiteit and at the academies of Antwerp and Brussels, is one of the key founders of the Art Nouveau movement, who, at age 25, fabricated his first domicile in Ghent just after finishing his studies at the Brussels Academy of Fine Art.

Belgium's extensive industrial development during that period, which was based on mining, iron and steel industries, led to the emergence of a new and well-off bourgeoisie that was readily disposed to exhibiting its recently acquired wealth and social status by commissioning original architectural creations. Industrial space had called for the development of large scale iron and glass constructions for factory use and this new technology became available for possible applications in creating human habitations. Grasping upon this new technology, Monsieur Horta broke the mould of traditional architecture (in search of the all-inclusive Gesamtkunstwerk) by giving his spaces a centralized light well.

The first Art Nouveau building was built in Brussels in 1893 by Horta, the Hôtel Tassel. The Hôtel Tassel, at 6, Rue Janson, was built for Horta's friend Emile Tassel as manifesto, and in so doing established Brussels as the capital of Art Nouveau.

Baron Horta's idea to construct lyrically enchanting spaces with whimsical arabesques (noodles, whiplashes and eels) is particularly evident in his 1900 *Maison Personnelle* (personal home) which is located at 23-25, rue Américaine, in Brussels. It is one of the most exquisite Art Nouveau buildings in the world and open to the public. Its fully immersive fin-de-siècle noise milieu is achieved through an unfamiliar warped suppleness of space created through thin, windblown, and whiplashed lines which lent me to the feelings of sprite underwater hair and, of course, writhing seaweed.

Baron Horta's creative fervor was enthusiastically received for almost 20 years (from 1893 to 1910) by an ample component of the Brussels' fin-de-siècle bourgeoisie. However, the Catholic high society rejected Art Nouveau, considering it dangerously decadent because of its emphatic use of curved lines. Consequently, in Brussels no church was ever built in the Art Nouveau style, which made it *ipso facto* a part of the secular movement of the period. Between 1915 and 1919 Horta stayed in England and the United States respectively, and there his whiplash noise ideal towards space became tempered and he turned to straight lines.

Gesamtkunstwerk-inspired designer/architect/theorist Henry Van de Velde (1863-1957) also was a key figure in the movement as he called for the unification of art into the space of the whole room (wallpapers, furniture and paintings). Van de Velde and his Brussels company, the Société Van de Velde, created all the interior furnishings of his buildings, including rugs and metalwork and in one case even a matching dress for its owner in keeping with his theories of totality which he articulated in *Déblaiement d'art* (1894) and *Aperçus en vue d'une synthèse d'art* (1895).

Van de Velde advocated in his tracts the unification of all of the arts as an instrument of social reform and a rejection of historical forms. Living in Germany, he became associated with the rise of the *Jugendstil* and became an early member of the *Deutsche Werkbund* (who invited him to build a theater for its planned exhibition in Köln in 1914). He is considerably known for his Havana Cigar Shop, a shop he created in Berlin in 1899 in collaboration with the Belgian painter/designer/theoretician Georges Lemmen (1865-1916). Lemmen was especially recognized for his carpets, wallpaper, and tiles. Indeed Henry Van de Velde's reappraisal of the status of the applied arts became a fundamental issue in the Sezessionist movement.

We turn now to the imposing suavity of Antoni Gaudí in Catalonia. At the age of 16, Antoni Gaudí (1852-1926) left his hometown Reus to join the school of architecture in Barcelona where he quickly adapted Islamic, Oriental and Gothic influences. Although he did not travel about Europe, Gaudí was acquainted with fin-de-siècle Belgium/French avant-garde movements because of the intimate relationship between Barcelona and France and with the pre-modernistic movements of Arts and Crafts, Gothic Revival, and Impressionism which were discussed in the intellectual proto-modernist circle which he frequented, but it was Horta's Art Nouveau movement that influenced Gaudí the most, stimulating him to experiment with new materials and new fluid shapes. Gaudí was particularly close with Count Güell, who travelled often in Europe and it was Güell who introduced Gaudí to the theories of the architect/theorist who exerted the most persuasive influence over the Art Nouveau architects in regards to the Gesamtkunstwerkkonzept ideal, Viollet-le-Duc and his book *Entretients sur l'Architecture* (which influenced both Horta and Gaudí). Noteworthy is the fact that King Ludwig paid a visit in 1867 to Pierrefonds, a restored medieval citadel that underwent restoration by Viollet-le-Duc.

Like Horta's, Gaudí's version of Art Nouveau noise, is characterized by an overwhelming proclivity for the organic nature of women, beasts, and plants which he translated into immersive utility. The materials utilized by Gaudí towards these ends ranged from stone, ceramic, tile, wrought-iron, glass and brick. He also used broken tiles for financial and technical reasons, as square tiles could not match the wavy shapes he preferred, plus it was cheaper to use free broken refuse from ceramic factories.

Antoni Gaudí is a chief exponent of Art Nouveau noise precisely with his 1906 building Casa Batlló located at 43, Passeig de Gràcia, Barcelona, noticeable for its organic tactility of bones and shells within, and its external cocked surf façade and chimerical roof. With Casa Batlló, Gaudí accomplished an astute transformation of an existing building, transforming it into an enchanting immersive Gesamtkunstwerk as Gaudí thoroughly undertook the design of every single element of the building, from the extravagantly protuberant façade to all aspects of the interior, including the gracefully gnarled furniture. On the exterior, Gaudí was able to combine a flamboyantly surging façade (in an ingeniously cool-color orchestration) while maintaining a dialogue with the neighboring Casa Ametller (1900), built by Josep Puig i Cadafalch (1869-1956) four years earlier. Powerful pillars that resemble the substantiality of mammoth elephant legs accost the visitor at street level, protruding into the sidewalk, nigh tripping up an unaware pedestrian. These legs are bordered by a craggy vertebrae-like tier and the wavy façade extends upward between these two biologically evoking forms, culminating at the roof in a gargoylesque humping crescendo. The façade itself, coated in a layer of Montjuïc stone, shimmers seductively under the sun in multifarious chameleon-like colors, fraught with a scattering of small roundish plates resembling fish or reptilian scales. Affixed to this seething mass of swelling construction are a number of small, elegantly curved balconies with oval shaped portholes.

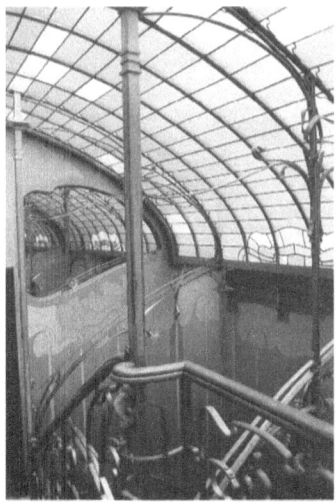

Figure 12 – Interior stairway of Victor Horta's *Hôtel Tassel* (Brussels)

The entire structure feels unsharpened, flowing and smooth in opposition to the street itself on which the arrangement sits, with the exception of a few square windows up top. Even the walls are gently rounded in strained undulation and contraction, as if they too have entered into the oceanic female throws of a fluttering uteral orgasm. The walls appear to be made of a soft, smooth, supple, leathery material and this illusion of softness is carried through by the roundness of the inside forms of the

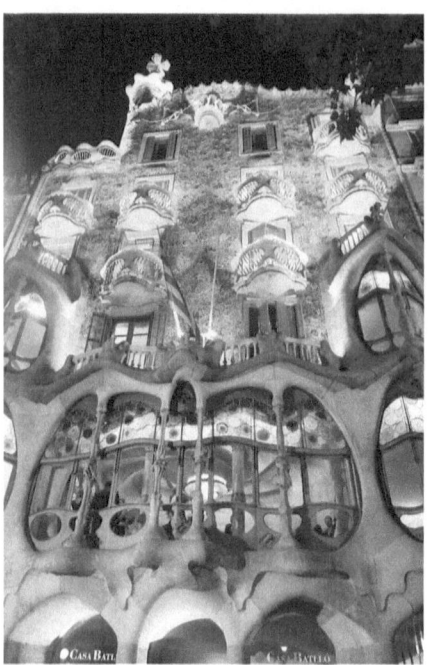

Figure 13 – Exterior view of Antoni Gaudí's *Casa Batlló* (Barcelona)

building where one has the feeling of being pleasantly encased in an expanse of hardened dripped honey. Turning, lunging stair railings are met, engulfed and supplemented by softly, heaving honey-colored walls and wooden biomorphicly shaped carved doors and irregularly shaped windows. There are no right angled corners or straight lines, which offers the immersant an impression of being wrapped up in one continuous fluid wave motion, complementary with the exterior. The number of ceramic tile elements used, that complement the feeling of inhabiting a construction produced by organic cells, increases towards the roof, where the crest of the roofing runs in a protuberant line that traces a zigzag spinal swell. The roof is covered in pallid bluish-pink ceramic tiles on the side facing the street, almost as if it were blushing due to the pithy sensuality of its avant-garde stance.

The French focus for Art Nouveau was Paris and the city of Nancy. In Nancy, one can still encounter a wonderfully complete Art Nouveau environment intact at the Musée de l'Ecole de Nancy's Salle à Manger (1904) by Eugène Vallin (1856-1936). Art Nouveau came to Paris principally by the celebrated architect Hector Guimard (1867-1942) who, as most people know, designed the Paris metro entrances, among other structures. In 1894 Guimard was building the Castel Beranger in a neo-gothic style when he visited Horta in Brussels. Inspired by what he had seen of the Hôtel Tassel, Guimard modified his plans for Castel Beranger, designing every detail, the wall paper, door handles, floor tiles, and front door, in Gesamtkunstwerk manner. Another architect who theorized, designed and built Gesamtkunstwerk-based Art Nouveau in Paris was

Frantz Jourdain (1847-1935), an influential teacher, theorist and builder of the 1910 La Samaritaine building.

Painting Modernist Noise

Today music, as it becomes continually more complicated, strives to amalgamate the most dissonant, strange and harsh sounds. In this way we come ever closer to noise-sound.
— Luigi Russolo, *The Art of Noises*

As initiated (by the students) in the studio of Gabriel-Charles Gleyre (1808-1874), Impressionist painting restored noise concentration onto the two-dimensional surface of the canvas while simultaneously suggesting an interregnum of luminous space. As such, it also played a consequential role in modelling fin-de-siècle French aesthetics. One thinks here of the ephemeral paintings of Claude Monet (1840-1926), particularly his extensive series of paintings, *Haystacks* (painted between 1890 and 1892) and *Rouen Cathedral* (painted between 1892 and 1894) and then *Nymphéas*, the series of 23 large paintings (19 of which were 2 by 4.30 meters (6.56 by 14 feet)) which he created late in life based on his Giverny garden's *Bassin des Nymphéas*. From 1915 to 1926 Monet exhibited all of these paintings in wall-to-wall installation mode, filling the three rooms of the l'Orangerie des Tuileries in Paris. But also indicative of this noise aesthetic are the paintings of Auguste Renoir (1841-1919), Alfred Sisley (1839-1899) and Jean-Frédéric Bazille (1841-1871).

Post-Impressionism extended this noise momentum, for example, with Georges Seurat's (1859-1891) mammoth 1886 Pointillist canvas *Un Dimanche après-midi à la Grand Jatte*. Here, everything on the canvas is inexorably locked together in one flowing noise as the composition (taken as a whole) postulates an uninterrupted enveloping energy devised from the color theory of the chemist Eugène Chenoiseeul (1786-1889). We can see this continue in the Divisionist paintings of Seurat's friend Paul Signac (1863-1935), for example in his painting *la Voile Jaune* of 1904 which shows a ship disintegrating into its environment. Paul Cézanne likewise extends this noise tradition in such a way, that his influence on the 20th century is hard to overstate.

As Rudolf de Lippe pointed out in his book *La Géometrisation de l'Homme en Europe à L'Epoque Moderne*, increasingly in the Modern era the geometricization of human vision became the general methodical condition in the West, characterized by an analytical sight which decomposes the immersive noise vision sphere into geometricized fragmented parts. This is a modern technological vision whose effectiveness lies in its tendency to isolate and decontextualize noise scope. Indeed, modern technology had an enormous social impact in the 20th century in this, and other, respects. The automobile and electric power, for instance, radically changed both the scale and the quality of 20th century life, promoting a process of rapid urbanization and a substantial change in lifestyle through mass production of household goods and appliances. The rapid development of the aeroplane, the cinema, and the radio made the world seem suddenly smaller and more accessible. Since 1900, the speed of travel has increased by a factor of 10 to the 2nd power, known energy resources by 10 to the 3rd, explosive power of weaponry by 10 to the 6th, and speed of communication by 10 to the 7th power. Such new ways of understanding involve a change in perspective, and that change is marked in the 20th century by an extended propensity for immersion into noise.

"An 'automatic' scribble of twisting and interlacing lines permits the germ of an idea in the unconscious to express, or at least suggest itself to consciousness. From this mass of procreative shapes, full of fallacy, a feeble embryo of an idea may be selected and trained by the artist to full growth and power. By these means, may the profoundest depths of memory be drawn upon and the springs of spiritual instinct tapped". So wrote Austin Osman Spare (1886-1956) in a short essay called "Notes on Automatic Drawing" in 1916 for the British art magazine *FORM*. This statement recalls the noise art in the Abside in Lascaux, but who the heck is Austin Osman Spare?

Spare was a chaos noise artist replete with potentials, but who has no place in the canon of Modern Art. He is a spiritual noise artist who concentrated on transforming his libidinal energy into noise art through the use of automatism almost ten years prior to the Surrealists. In a way, he is an artist with whom I cannot be satisfied, but with whom I can be impressed in terms of the art of noise. Why impressed? Because among the many complexities that have transpired in today's society due to the delirious effects of information-communication technology—and the proliferation of visual information that has resulted from this technology—is the changing nature of artistic definition. And Spare's use of automatic instinct in creating his noise art addresses this condition fully. As you may know, automatism, in the arts, is an act of creation which either allows chance to play a major role or which draws on the unconscious mind through free association, states of trance, or dreams. Spare was a pioneer in this noise practice specifically with his experiments in trance, which is basically self-initiated work with reflexive feedback loops—the basis of cybernetics.

He is impressive, too, in philosophical terms, as contemporary postmodern thought has been concerned with the poststructuralist deliberation on the notion of the subject in order to question (and unlasso) its traditionally privileged epistemological status. Particularly in respect to the automatic-assisted techno-artist (an artist whose discourse revolves around networks and rhizomes), there has been a sustained effort to question the role of the artist/subject as the intending and knowing autonomous creator of art—as its coherent originator. Again Spare's automatism informs us here. In fact, for me, the semi-automatic drawings of A. O. Spare have become emblematic of this question of the rigorous scrutiny of the subject which Jacques Derrida (1930-2004) has described of as logocentrism: the once-held distinctions between subjectivity and objectivity; between public and private; between fantasy and reality, and between the unconscious and the conscious realms.

Today, we understand that these distinctions are breaking down under the pressure of our speeding and omnipresent computer communications network technologies. We are now part of an automated technologically hallucinogenic culture that functions along the lines of a dream, free from some of the strictures of time and space; free from some of our traditional earthly limits that have been broken down by the instantaneous nature of electronic communications.

The modernist existential concept of the singular individual has been supplanted by the electronic-aided individual, in a way liberating her from linear time, and vaporously placing her in a technologically stored eternity (simulacrum-hyperreality). This quality of phantasmagorical and perverse displacement has for some signified a tightening spiral which formulates a new vision of existence, a vision which Jean Baudrillard has called *pornographic* and which Deleuze and Guattari have called *schizoid*.

Both these descriptions apply aptly to the drawings of A.O. Spare in a variety of ways that I will make apparent shortly. For those, and they are numerous, who are not familiar with the work of Spare, let me first provide some rudimentary background on him.

Austin Osman Spare was born the son of a London policeman. Doom loomed abundantly in Fin de Siècle England as Spare came of age, thus his development into what can now be recognized as a late-decadent, perversely ornamental, graphic dandy in the manner of Felicien Rops (1833-1898) and/or Aubrey Beardsley (1872-1898) can be readily contextualized.

As a young man, Spare was for a brief period of time a member of the *Silver Star*, Alister Crowley's (1875-1947) magical order. Spare's lifelong interest in the theory and practice of sorcery was initiated, he recounted, by his sexual relationship at a very young age with an elderly woman named Paterson. To perform sorcery, for Spare, was a practice meant to captivate, encircle and ensnare spirits. It is not quite the same thing as practicing magic, which is the art of casting spells or glamors. For Spare, as well as for Crowley, Tantricesque sex—the withholding of the orgasmic—held the means of access to their magical systems. However, it is in Spare's conception of radical and total pan-sexual freedom, consisting in the unrestricted expression of what he held to be the "inherent dream", where we first detect the seditious and chaotic philosophy which drove a prong between himself and Crowley—and every other esoteric system but his own brand of chaos magic/art.[240]

In 1905, at the tender age of 17, Spare self-published his first collection of drawings in a book of aphorisms entitled *EARTH INFERNO*. In it, he lamented the death of what he called the "ubiquitous women of unconsciousness" (he believed that out of the flesh of our mothers come dreams and memories of the gods), and castigated what he called the "inferno of the normal". For Spare, and I agree with him here, there are no *levels* or *layers* to consciousness, and no dichotomy between the *conscious* and the *unconscious*. There isn't even a clearly definable boundary between *consciousness* and the *object of consciousness*, between *subject* and *object*, between *action* and *situation*. There is only a depth or thickness of consciousness which varies in proportion to our state of self-awareness—from the thinnest film of near being, where we engage in pure desire/instinct driven towards action, to so paralyzingly thick opacity that it induces catatonia. The point of automatism is that the more spontaneously we act, the less self-conscious we are.

EARTH INFERNO disparages the world of humdrum banality in favor of an exotic pan-sexual orb which Spare began to reveal in a spate of non-automatic drawings somewhat reminiscent of the decadent artists previously mentioned. His intention was pan-sexual, transcendental, and androgynous—in that Spare claimed that he was all sex—and that what he was *not*, was moral thought: simulating and separating. Moreover, he wrote that when belief detaches itself from the accessories of convention, desire stands revealed as the ecstasy of the self, ungoverned by its simulated forms.

In 1907, Spare self-published a second collection of drawings in a publication named *THE BOOK OF SATYRS* which contained acute insights into the social order of his day. Then in 1909, Spare began work on a third book, this time of semi-automatic drawings entitled *THE BOOK OF PLEASURES* on which he worked for four years. This book emerged in 1913, as did another called *THE PSYCHOLOGY OF ECSTASY*. In 1914, he held his first one-person exhibition at the Baillie Gallery in London. It included many of the semi-automatic sketches he drew while half asleep or in a self-induced masturbatory trance. Indeed, most of Spare's semi-automatic works (from 1910 onward) were produced in onanistic self-induced trances that he claimed were sometimes controlled by intrusive occult intelligences working through him. Here, through masturbatory trance, he said the "I" becomes atmospheric. This certainly reminds us of the disembodied state so often encountered in electronic environments (such as virtual reality) where the so-called *self* is uncoupled from the body and pseudo projected into computerized space.[241] Indeed, Spare considered his best accomplishments those which he said were produced through him by disembodied spirits rather than by him, often by the hand of the phantom spirits of William Blake (1757-1827), Leonardo da Vinci, Hans Holbein the Younger (1497-1543) and Albrecht Dürer (1471-1528). Not bad virtual company. Spare quite wildly would declare that his was the automatic hand utilized by these deceased masters. Through his automatic and delirious technique, Spare claimed to be able to draw upon the profoundest depths of memory and to tap into the springs of instinct. These drawings can be found in a book Spare prepared but never self-published in 1925 that he called *A BOOK OF AUTOMATIC DRAW-*

INGS, a book that was posthumously published in 1972. More automatic drawings were lost when on May 10th, 1941, during the height of the London bombings, Spare's London flat was obliterated by a bomb.[242]

It is in his highly extravagant practice of automatic openness and swank self-denial/self-pleasure that Spare's relevance to the poststructuralist/post-Internet conceptions of the noisily decentered and distributed subject is found. Specifically, Spare's heterogeneity relevance here lies in his interests in the loss of subjectivity as experienced in sexual transport and sexual fantasies, interests which now dovetail into our interests in the philosophical loss of sovereignty typical of the art of noise. Here, for example, with the loss of body consciousness specific to total-immersion within a virtual reality environment, one frequently senses a transporting dissolution moving consciousness away from self-consciousness.[243] Also there is an obvious bearing on aspects of on-line faux self-permutations—what are called *avatars*. By participating whole-heartedly in his insertion (and semi-fake disappearance) into the transpersonal symbolic economy of the sign through the assumed equivalence of life and death (in what perhaps can be imagined for us as digitized-stored post-existence), Spare remains truly an individual, if not altogether alone in his time. His was a radical transcendentally false egoless gesture (what a bogus collaboration!) which he fabricated in order to make semi-automatic art try to do magical things. In the process, he created an exciting conception of noise art which focuses on collective and collected selves. Undoubtedly, his is a view which counters the long-standing western-metaphysical-phallocratic-heroic portrayal of male-selfhood—a view which we all know too well. And yet, doesn't his view of a compiled self, akin to the essence of the death of the subject, offer just the sort of resistance to the structures of logocentric civilization that simulationist theory claimed was impossible?

Listen to what Spare wrote on this in a 1916 essay: "Let it not be thought that a person not an artist may by these means become one: but those artists who are hampered in expression, who feel limited by the hard conventions of the day and wish for freedom, these may find in automatic drawing a power and a liberty elsewhere undiscoverable".[244]

Spare's quite early conception of the illusory coherence of the "I", renders everyone and every-sex equally phantasmagorical (as disembodied fabula) akin to the way the speeding electronic-computer network can. His conception of automatic every-sex was clarified when he wrote, "In the ecstatic condition the mind elevates all sexual powers towards infinity". And, "Speed is the criterion of the genuine automatic. Art becomes, by this velocity an ecstatic power expressing in a metaphorical language the desire for joy". But, in effect, his pan-sexual joyful "I" existed primarily as the construct of a system of male forces which he claimed acted through him on the creation of a synergistic complex image. This synergistic compounding of the mnemonic threshold encapsulates our current post-postmodern-networked predicament in that the fabulated digital-self today may feel sublimated by the automatic system in which it operates. It may feel eclipsed—but also freed-up by—the mammoth computer-media-web as phantom information bits flow continuously around and through us in a vague endless whirl of unverifiablity. This digital-self unquestionably partakes in a data proliferation which forms, bit by bit, into an extensive aggregate somewhere deep in the abstruse recesses of our hard drives, a data proliferation that is awaiting discharge and reformation through noise art.

Perhaps by automatically stirring the viractual-self, Spare can be understood as a precursor of digital fluidity/copy-ability, working as he did, vis-à-vis onanistic actions while forestalling the actualization of his orgasm—thus maintaining an extended virtual state of self-pleasure. Certainly his remarkable sex/magical method for making noise art suggests a methodology based on obsession and longed for ecstasy which I have taken as my digital working method too—a method that plays in the area of control/non-control with an aim towards constructing a capricious alliance that associates discourses of machinic noise with organic sexuality, an association which opens up both notions to mental connections that enlarge them. The digital-noise-self here is impregnated by a sustained desire that becomes energized by the supposition that deep memory responds to chaotic longings and can relive original obsessions. In relationship to this method, Spare said, "The artist must be trained to work freely and without control within a continuous line and without afterthought—that is, the artist's intentions should just escape consciousness. In time, shapes will be found to evolve, suggesting conceptions, forms—and ultimately style".

So it is extremely relevant, then, to consider Spare's means of becoming courageously individual through his frenzied tranced-groupings. In effect, he achieved this through the transgression of (and by!) his artistic "masters". In terms of the original's unimportance to our electronic era's conception of art as simulation, Spare's claim to meta-individuality in his production (really what he claimed was a co-production achieved through automatic means) seems prophetic. If a substance-less collective history of digitized art images and the unseen labor of computer programmers lurks and reverberates internally in each technologically-aided art work today, and if in each of our computers a data-bank of visual information lingers beyond our personal propensity and (perhaps) dominates us, then an inner freedom from external authority indeed seems futile. We can only act with what authority has passed down to us. But what if the search for a digitally-assisted noise art in our contemporary context of the information society is more simply directed towards not repeating what has been learned and collected? Perhaps this possibility—as achieved through the automatic unconscious act—is what I have chiefly learned from Spare's work and writings, as well as his exclusion from the canon of art history. By way of semi-automatic processes, art can be further problematized, cracked-open, drained and transfigured through the strange mixture Spare showed us of disinterested rapture—a generous elation where off-beat panoramas and chaotic multiple personalities have room to emerge.

To achieve this, Spare would first exhaust himself before beginning to draw in a somber candle-lit room and in a slight trance with no particular idea in mind, thereby, he believed, reaching deeper and more remote layers of chaotic memory. He did this while continuously abhorring the accepted values and maudlin conceits of his day. It has been my experience that computer programming that utilizes automatic functions can achieve similar ends. I learned this through developing in 1991-2 a real-time, operative artificial-life application based on the viral model.[245] This disruptive model, though based upon nature, makes use of automatic functions of computation to circumvent conscious control. Such a non-rational, unpredictable automation, of course, stands in stark contrast to the automation of Fordist/Taylorist production, with its legacy of instrumental rationality.

The fact that Spare was a sensual occultist should not misdirect our appreciation of his artistic and theoretical endeavor. The logic of noise facilitated by the Internet is satiated with a parallel concealment. For most, there is much mystery attached to the digital hidden codes and routing formats that expedite our tele-communications. Moreover, his drowsy semiautomatic drawings, with their multifarious and allusive search for something antithetical to the established norm—and their morbid subversion of the concept of individuality and authorship—play well on today's desire for excessive noise art that the computer tends to encourage. Spare's drawings enmesh, hinder, alter and disrupt the mundanity of elementary communications with their inexorably chimerical noise.

Today it is in the hyper-logic of the endlessly duplicable digital noise and/or sound where we can probe, much as Spare did, for a particular and personal occult expression. Also, we should remember that, within the current electronic environment of hypermedia, artistic annihilations of linear time are now possible. Thus barriers between the deceased and living become somewhat abolished. This, too, recalls Spare's chaotic methodology. In his own fashion, he created a non-linear noise where deep-memory threatens the common order of events, thus questioning both clichéd ideas of originality and supplied social codes. Clearly, with its emphasis on origin, author and finality, his non-linear artfulness subverts the modernistic conception of production, but without merely accepting the artificial, the copy, the simulation, as the end point.

This is how the art of noise functions in our technomediacratic society—a hyper-society that deploys the effects of rhizomatic connections and trance-like repetitions. It is the artist's task today, I feel, to disadvantage the digital reproductive technology so as to defeat its attempt at negating our art's mystical significance. But to do this, we must abandon the Enlightenment baggage of authorizing categories and live non-linearly, while accepting nothing as flatly given. Here again, Spare inspires as he explicitly eschewed categorization and instead sought to problematize the authority of the category through hyper-logic. So Spare compels us again to take notice of the various ways artistic conventions have molded our responses and regulated our artistic denotations.

The possibilities of non-linear, complex-entangled-erotic noise configurations springing forth from the digital Id made up of mercurial symbols and pan-sexual concepts in opposition to recycled representations provides an interesting insight into the way Spare's art (with its convoluted compositions made up of vague confiscations) directs us towards the conception of the transformative possibilities of technologically-aided noise art.

Perhaps the hope that Spare's non-linear and semi-automatic noise art can show us a way to resist art history's drive towards reification is a fragile hope indeed in our electronically-homogenized cyberage. Honestly, such a hope may be less than we deserve, but it also may be more than we usually allow ourselves to envision.

What I am certain of is the need for spontaneous, pre-rational actions in the realm of art and technology so as to pursue spiritual and erotic desires, and here Spare inspires as he indicates ways in which we may escape the prison of technological logic to encounter intimate realities bound only by the next thought and driven only by the last. This is the answer to the question "how shall I be free today?" and to best express free thought through noise art without too many tainted preconceptions.

Not to dismiss Austin Osman Spare (and his concept of the collective self—which for us can be reconceived of as technological hyper-thought) as dilettante folly is to become aware of the fact that underlying everything virtual is a web of hyper-connections upon which we can exert more manipulative desire than we are normally led to believe by the society of the spectacle. But to do so, we must actively use noise art and not be content with merely consuming it. For as Félix Guattari said in his noteworthy book, *Chaosmosis: An Ethico-Aesthetic Paradigm*, the work of art, for those who use it, is an activity of unframing, of rupturing sense, of baroque proliferation or extreme impoverishment that leads to a recreation and a reinvention of the subject itself.

This reinvention of the self occurs through a curious alliance between the cold impersonality of technology and the flames of personal ecstasy in the new noise art of our time.

However, modernism in architecture (and then art) rejected such late Art Nouveau noise ideals by placing emphasis on the *unity and similarity of reductivist forms*. This reductive urge is called orthodox *Modernism*, an ideal concerned with essences and abstractions. Charles Jencks takes the view that the modern movement in architecture is based on a world-view informed by the Industrial Age with great emphasis on the mass production of virtually identical goods, and with reducing the design of something to its simplest functional elements. Simultaneously, profoundly new noise concepts in art began to appear in 1905 with Henri Matisse (1869-1954) and the Fauves at the Salon d'Automne in Paris. In particular, the Cubism of Pablo Picasso (1881-1973) and George Braque

(1882-1963) emerged as a radical departure from the perspectivist representational tradition of the past, as Cubism aimed to restructure representation through a redefinition of realism. *Analytic Cubism* (1908-1912) dropped the conventions of Renaissance framing in favor of a multi-outlook exploration of many different angles and viewpoints, articulated through overlapping and interlocking planes, as we see with Picasso's 1910 canvas *Portrait of Daniel-Henry Kahnweiler*.

In opposition to point-perspective, Analytic Cubism shows that the viewer synthesizes fragmentary accumulated evidence into an assembled totality when the viewer's volume of perceptions are detected through ambient vision in motion. Analytic Cubism re-analyzes and synthesizes vision's multiple viewpoints concurrently by tenaciously folding them (simultaneously) into one sweeping but minced formation. As such, Analytic Cubist consciousness suggests an embedded hermeneutic immersion into noise. *Synthetic Cubism* emphasizes this non-illusionistic program in a broader way through the incorporation of elements in the environment, such as fragments of wall-paper, journals and/or photographs. A good example is the 1912 legendary, *Nature Morte avec Chaise Canée* (Still-life with Chair Caning) by Pablo Picasso.

The early photomontages by the Dada artist Hannah Höch are particularly relevant to noise vision, for example with her amazing *Cut with the Kitchen Knife Dada through the Last Weimar Beer-Belly Cultural Epoch of Germany* from 1919-20.

In this work, Höch presents us with a rush of all-over and intermingled visual fragments (like in the abside of Lascaux) that exceeds any attempt at a clean, clear signal reception. As I showed in Lascaux, this form of visual noise presentational excess offers up the possibility of multiple interpretations that may be in conflict with each other. Thus the interpretive act seems to have no end here.[246]

So now we can turn fully to the immersive work of the Dadaist artist/designer/typographer/poet Kurt Schwitters (1887-1948) and his Hannover *Merzbau* (1923-1937), as stated a large influence on the Japanese noise master Merzbow. In the early-1920s, Schwitters began working on a collage/column which he called a *Schwitters-Säule* which soon grew out and up over the ceiling of his apartment in Hannover.[247] Soon it grew down across the walls, and niches were made in it to contain mementos from his friends which were later covered over until the work finally grew up through the ceiling, down through the floor and onto a small projected roof. In its entirety it was called *Merzbau* (Merz-house).[248] This Merzbau was abandoned to the Nazis when Schwitters moved to Norway to escape them. There he began another Merzbau (the Hus am Bakken at Lysaker) which was burnt down by children in 1951. The Hanover Merzbau, too, had been destroyed in the aerial Allied bombings of 1943, but in 1947 Schwitters began work on his final piece of what he called *total art*[249] —his *Merzbarn*. This work was to be made almost entirely of plaster with found objects embedded in it. Another relevant Schwitters project from the immersive perspective was his theorization of the *Merzbühne*, a "total-Merz-theater".[250] Though this project was never realized, it paralleled a number of other *total theater* projects that were developing in Europe during the 1930s.

In light of Schwitters' achievements, we might now consider Clarence Schmidt's 1930s noise decor/assemblage creation of a continuous chain of grottoes and corridors and caves created on O'Hayo Mountain near Woodstock, New York, that Allan Kaprow wrote about in his 1966 book *Assemblages, Environments, Happenings*.[251] Schmidt's collage grotto/labyrinth has been hailed by Adrian Henri as "possibly the 20th century's finest piece of *total art*"[252] —a concept of *environmental art* Henri developed in his book *Total Art: Environments, Happenings, and Performance* which, as we have seen, stems from the Wagnerian terminology Gesamtkunstwerk: a coextensive configuration which sets out to inexorably dominate, overwhelm, and flood us with sensory impressions.

The key European avant-garde movement with initial immersive art noise suggestion was Futurism.[253] I have already pointed out the crucial importance of Luigi Russolo's work in this respect and his 1913 manifesto *L'Arte dei Rumori* (translated as *The Art of Noises*) is seminal. The noise work of Fortunato Depero is also outstanding, such as his 1915 *Edificio di stile rumorista transformabile* (*Building of transformable noise style*)—in collaboration with Giacomo Balla[254] —and his *Anihccam del 3000: Canzone rumorista* (*Machine of 3000: Noise song*) (1916-24).

The Cubist ontological embeddedness of the view into a spread of moving optical fields was amplified in Italian Futurism, as it attempted to coalesce and condense scattered/totalized ocular impressions. Umberto Boccioni's (1882-1916) 1911 painting *States of Mind II: Those Who Go* is an admirable example. Responding to the machine age, the Futurists, under the philosophical leadership of Filippo Tommaso Marinetti (1876-1944), glorified speed and the machine while expressing a rejection of the past, best exemplified by a famous fist-fight in 1910 between the Futurist painters and poets and the Venetian townspeople who reacted in anger when 800,000 manifestos, entitled *Against Past-Loving Venice*, were scattered upon them.[255]

Cubo/Futurism achieved a syntheses of the flickering noise optics of Post-Impressionism with the spread of urban media visual production, announcing the postulate that reality is discovered through the slant of drifting involvement as opposed to static detached understanding. This dematerialized optical noise awareness suggested further supra-visual reconfigurations that are picked up by the radical avant-garde of mid-20th century, as we will see.

But rather than coming from Cubo/Futurism, art conceived of as *total experience*[256] stems, according to Henri in his book *Tota Art,* from Dada, a reaction against the First World War of 1914-1918.[257] It is true that the Dadaists did not restrict themselves to being painters, writers, dancers, or musicians, as most of them were involved in several art forms and in breaking down the boundaries that kept the arts distinct from one another. In particular, Henri suggests that *total experience* stems from Max Ernst's first Köln exhibition in 1920, in which Ernst was joined by other artists who, like him, were later to become Surrealists. The exhibition was entered through a men's urinal that was opened by an adolescent girl in a First Communion dress reciting obscene verses.[258]

Freud's intrigue with the unconscious was enthusiastically taken up by the Surrealists who saw his studies of dreams as central to their own desire to disrupt the norms of conscious perception. As Henri Ellenberger's book, *The Discovery of the Unconscious: The History and Evolution of Dynamic Psychiatry*, forcefully demonstrates, Freud did not "discover" the unconscious —if we can say that anyone did, it would be Jean-Martin Charcot (1825-1893) and Pierre Janet (1859-1947), the author of *De l'Angoisse à l'Extase* (Of Anguish and Ecstasy) —but Freud, working with his associate Josef Breuer (1842-1925), might be said to have posited the general principles and contents of the unconscious mind that gained predominance in the 20th century.

Henri states that the grand Surrealist exhibitions of 1936 in London and 1938 in Paris are the most direct precursors of total art. In the Paris show, under the direction of Marcel Duchamp,[259] the ceiling of the main room was hung with 1,200 coal sacks filled with paper while a gramophone played German military marches, complimented by an ornamental pool and the smell of roasting coffee-beans. For a later exhibition in 1942 at the New York Reid Mansion, entitled *First Papers of Surrealism*, Duchamp created an environment out of kilometers of entangled string.[260]

We must, too, acknowledge and indeed honor an immersive noise masterpiece and source of immense inspiration to the Surrealist movement itself, the *Palais Idéal* of Ferdinand Cheval (1836-1924) which, to my eye, appeared to be one gigantic sprawling noise grotto when I visited it, as if the stupendous mannerist grotto-façade at villa Borromeo had been left to grow untrimmed and run amok. It was constructed by Cheval alone in the Hauterives (Drôme) between the years 1879 and 1912, the result of 93,000 man hours of hard labor.

So a question: why does quietly framed pictorial art become progressively challenged by visual noise—and to a certain extent, eclipsed by it—following the Second World War?[261] Evidently there was something endemic in the barbarous conditions of 20th century modern warfare that facilitated this noise development at its onset, rather than any more laudable human aspirations towards the expanding of aesthetic perceptual consciousness. We can find examples of cultural visual noise previous to the war on occasion, as we have seen, but after it there is an explosion.

I have deduced that something in the consciousness of society was altered following the war and have further deduced that the bombing of civilian centers in the course of the war (that is, Köln, London, Tokyo) culminating with the American atomic bombings of the civilian Japanese cities, Hiroshima on August 6th, 1945 (circa 140,000 victims) and Nagasaki on August 9th, 1945 (circa 70,000 victims), changed the world's sense of space radically.

The Allies' strategic air offensive against Germany began to attain its maximum effectiveness in the opening months of 1944. Both the U.S. air forces concerned, namely the 8th in England and the 15th in Italy, were increased in numbers and improved in technical proficiency. By the end of 1943, the 8th Bomber Command alone could mount attacks of 700 planes and, early in 1944, regular 1,000 bomber plane missions became possible. Even more important was the arrival in Europe of effective long-range fighters, chief of which was the P-51 Mustang.

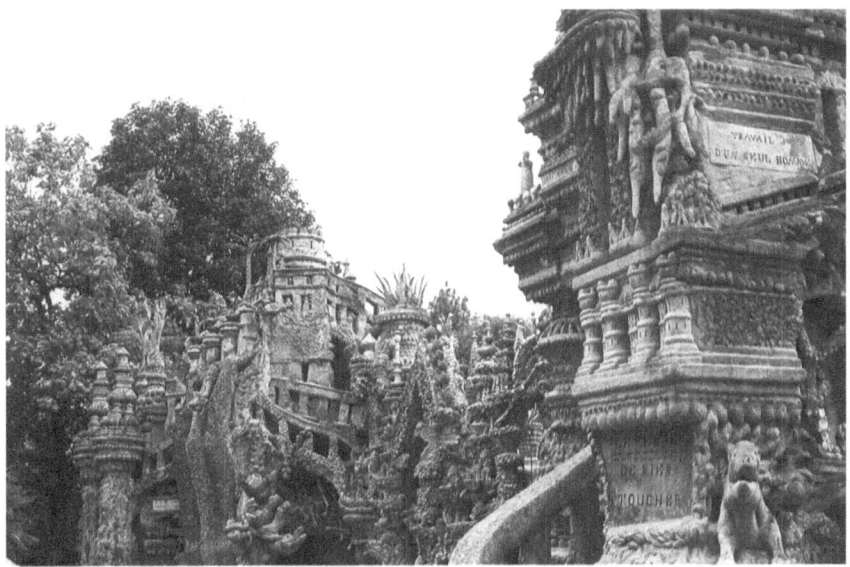

Figure 14 – Exterior view of the *Palais Idéal* by Facteur Cheval (Hauterives, Drôme, France)

However, Paul Virilio, in his esteemed *Bunker Archaeology*, indirectly suggested the initial date of this spatial consciousness transition as 1943, with the Nazi preparation for the first operational launching of the V-2 ballistic missile. Ballistic missiles are rocket-propelled weapons that travel by momentum in a high, arcing trajectory after they have been launched into flight by a brief burst of power. Although experiments had been undertaken before World War II on crude prototypes of the cruise and ballistic missiles, modern weapons are generally considered to have their true origins in the V-1 and V-2 missiles launched by Germany in 1944 and 1945. Both of those *Vergeltungswaffen* (Vengeance Weapons) defined the problems of propulsion and guidance that have continued ever since to shape cruise and ballistic missile development. Indeed strategic missiles represent a logical step in the attempt to attack enemy forces at a distance. As such, they can be seen as extensions of either artillery (in the case of ballistic missiles) or manned aircraft (in the case of cruise missiles).

In 1944, at the Peenemünde base on the island of Usedom in the Baltic, Wernher von Braun and his team created the V-2. The V-2 was 14.1 meters long (47 feet) and its payload was about 900 kg of high explosives. The horizontal range was about 350 kilometers (220 miles), and the peak altitude usually reached was about 100 kilometers (62 miles).

It was first fired against Paris on September 6th, 1944. Two days later, the first of more than 1,300 V-2s was fired against Great Britain (the last on March 27th, 1945). Belgium was bombarded almost as heavily with them. Reaching a height of more than 160 kilometers (100 miles), the V-2 marked the beginning of the space age. After the war, both the United States and the Soviet Union captured large numbers of V-2s and used them in research that led to the development of their missile programs.

Nevertheless, Pablo Picasso's 1937 monumental 3.51 by 7.52 meter (11.5 by 24.6 feet) painting *Guernica* presented into art consciousness an earlier (the first) civilian air-bombardment of innocent people at home in their city of Guernica Y Luno during the Spanish Civil War (1936-1939). Here 1,654 Basque people were killed, at the bequest of Francisco Franco Bahamonde (1892-1975), and 889 were wounded, including the elderly, women, and children by Adolf Hitler's (1889-1945) Junker 52 and Heinkel 51 warplanes in the service of Spanish fascism.

Previously, there existed a separation between military and habitational space but, with the bombing of Guernica Y Luno, the swathed immersive space of the tellurian domain was suddenly deemed defunct as previous earth/covering frontiers became increasingly porous to airborne invasions. This sense of airborne vulnerability soon extended itself further and further outwards with the launching of spy and then military-communications satellites (Sputnik in 1957), the first manned space flight of the Soviet military-pilot Yuri Gagarine (1934-1968) on April 12th, 1961 (the first man in space), and then the first manned trip to the moon of the American Apollo Mission in 1969 which featured Neil Armstrong's televised trek on the moon. Rocket technology enabled military forces to put nuclear weapons on intercontinental missiles, due largely to the former work of Russian rocket pioneer Konstantin Tsiolkovsky (1857-1935) (whose visionary ideals came from Nikolai Fedorovich Fedorov (1828-1903)), the American Robert H. Goddard (1882-1945) and the German Hermann Oberth (1894-1989). With rocket technology, the space of military interaction clearly expanded and, mirror-like, entered the inner dimensions of the human psyche. Virilio verifies this shift in consciousness in his book, *War and Cinema: The Logistics of Perception*, where he traces the colonization of the unhurried gaze by military technologies and the introduction of military intelligence into the indoctrination of the non-combatant's perceptions. This "rational" scopic extension of vision is accomplished precisely at the loss of another sort of vision, the ambient/holonogic, as it involves a heightened ordering and sighting of linear perspectives and a consequent geometrization of both external space and the inner human. This new sense of threatening external space perhaps is most strongly, and most fearfully, exemplified by what has become known as *C3I* (pronounced as *see cubed eye*), the electronic military intelligence spatial fusion of control, command, communication and intelligence that developed as the electronic/digital system of strategic command over the U.S. military's nuclear arsenal. Herbert York, in his essay "Nuclear Deterrence and the Military Uses of Space", provides a fine short overview of this trend towards militarizing and sighting outer (and hence inner) space where he outlines the Strategic Defense Initiative (SDI) program of the 1980s and its ensuing militarization of outer space. Indeed, York makes the point that, "from the beginning", the use of the space program has been "primarily of a military, not civilian or scientific nature".[262] In 1983, as part of the SDI program, President Ronald Reagan put forth his "vision" of what became pejoratively called *Star Wars*, perhaps the archetype of this oppressive spatial consciousness.

Certainly, it is true that, hidden in the computer, there is something so strong, so repetitious, so ominous, and so pregnant with the darkness of infinite noise that it excites and frightens us. This is why the innumerable ramifications of mechanical desire help us to utilize our unconscious mind. And this is the real answer to why computers are interesting in art. We admire their inhuman beauty. They return us to the experimental, to a state of sexual desire and noisy restlessness. The neural processes they mimic are our own deepest desires and meticulous obsessions. Their repetitions are the fusion repetitions of our sexual acts with their duplication of eggs, sperm and blood.[263]

Of course, nuclear weapons derive their enormous explosive force from either the fission or fusion of atomic nuclei. Their significance may best be appreciated by the coining of the words *kiloton* (1,000 tons) and *megaton* (one million tons) to describe their blast effect in equivalent weights of TNT. For example, the first nuclear fission bomb, the one dropped on Hiroshima, Japan, in 1945, released energy equalling 15,000 tons (15 kilotons) of chemical explosive from less than 130 pounds (60 kilograms) of uranium. Fusion bombs, on the other hand, have given yields of up to almost 60 megatons.

The first nuclear weapons were bombs delivered by aircraft. Warheads for strategic ballistic missiles, however, have become by far the most important nuclear weapons. The U.S. stockpile of nuclear weapons, which included the hydrogen bomb that was first test exploded in 1952, reached its peak in 1967 with more than 32,000 warheads of 30 different types. The Soviet stockpile reached its peak of about 33,000 warheads in 1988. Throughout the ballistic missile arms race, the United States tended to streamline its weapons, seeking greater accuracy and lower explosive power, or yield. Most U.S. systems carried warheads of less than one megaton, with the largest being the nine-megaton Titan II, in service from 1963 through 1987. Meanwhile, the Soviet Union, perhaps to make up for its difficulties in solving guidance problems, concentrated on larger missiles and higher yields. The Soviet warheads often exceeded five megatons, with the largest being a 20 to 25-megaton warhead deployed on the SS-7 Saddler from 1961 to 1980 and a 25-megaton warhead on the SS-9 Scarp, deployed from 1967 to 1982. Hence at mid-20th century, space became the range of both humanity's greatest fears (nuclear extinction of life on the planet) and its boldest aspirations (co-operative peaceful space exploration).

What I am proposing here, in agreement with Virilio, is that the sense of human-enfolded space was radically transformed in 1943 when the German rocket-launched bombs began to fall on London without warning, shattering the common sense of civilized, non-combatant, protected space, and that this remade human feelings towards external space thoroughly. As a consequence, I maintain, a consciousness of civilian aerial bombing, of atomic weapons, of military rocketry and of the eventual militarization of outer-space has greatly engendered the abandonment of the horizontal line in art, which for thousands of years had been the basis of aesthetics and proportion. Of course accompanying this new sense of noise space was a general post-war urge to position one's artistic activities and ideas outside of previous contexts; in western art and philosophy's case, outside of Surrealism and Existentialism.

In terms of a transformation of our sense of internal space, I find it amazing that Dr. Albert Hofmann (a biochemist at the Sandoz pharmaceutical firm in Basel, Switzerland) accidentally discovered LSD (lysergic acid diethylamide tartrate) the same year that rocket-launched bombs began to drop from the sky: 1943. LSD was first synthesized in 1938

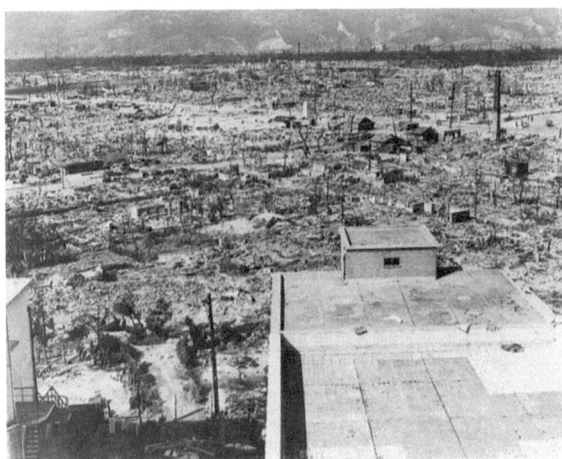

Figure 15 – Hiroshima after the dropping of the atomic bomb, 1945

by Hofmann but he did not know what he had synthesized until 1943 when he accidentally absorbed a small amount of LSD (which is colorless, odorless, and tasteless) and thus discovered its visionary properties. With this ingestion, Hoffman, after surveying the room he was in, realized that he now formed a nice noise continuum with everything in sight. The room seemed to shimmer in the sunlight, and he became aware of the atomic substructure that underlay the visible world of the senses.

The problems of LSD's experiential description are notorious, and the typology of its effects vary, but the central experience is one where a new level of consciousness emerges. As this cultural phenomenon did much to change the art of the 1960s to the 1980s, I shall attempt to describe LSD's salient properties as they apply to the art noise experience. Foremost in this regard is that, when experiencing the chemical, the awareness of individual identity somewhat evaporates and subject/object relationships tend to dissolve. The world seems as if it is simply a fluid, shifting extension of mind and it shimmers as if it were charged with a high-voltage electricity. Additionally, the subject often feels melted into the environment and somehow contiguous with it and there is an acute awareness of the atomic substructure of reality which makes it seem that one could pass through a wall or another person. Most importantly, the subject is somehow united with a sense of *unified ground of being*, and that urge, as we have seen, has driven the Gesamtkunstwerk ideal since the beginning of time.

As Bohm and others have shown us, everything in the universe is made up of, and seen as, part of the seamless *unified ground* of the holographic total-fabric, and LSD seems to make this visible. Furthermore, this *unified ground paradigm* began forcefully entering Western consciousness just following World War II's brutal demonstration of nuclear destructive power on Japan and, as we will soon see here, began to be reflected forcefully in vanguard art of the post-war period.

The Noisiness of Jackson Pollock

As reported in the artist Raphael Soyer's (1899-1987) *New York Times'* obituary on November 25, 1987, Jackson Pollock (1912-1956) said to Soyer one day: "Why do you paint like you do? There are planes flying, and you paint realistically. You don't belong to our time". From this statement, we know that even though Pollock took his first trip on an airplane (to Chicago) in 1951, he was obviously acutely aware of the expanding nature of technological space that was defining the 20th century and its art. Pollock maintained that paramount concern in a radio interview with William Wright in 1950 when he said, "My opinion is that new needs need new techniques. The modern painter cannot express his age, the airplane, the atom bomb, the radio in the old forms of the Renaissance". According to Pollock, if art is to be contemporary, by definition, it must address the issues of its time.

The impact of Picasso's *Guernica* (which arrived in New York City in 1939 at the Museum of Modern Art) with its mural size tied to its noisy theme of aerial bombardment on a defenseless population had a stirring effect on Pollock. *Guernica*'s inventive formal automatism (which had been introduced to Picasso via the French Surrealists) and its social conscience, in conjunction with its exploration of collective unconscious fear and guilt, served as a dramatic catalyst for Pollock's expanding idea of art that culminated in the great all-over syncretistic paintings beginning in 1947.

In the late-1940s and early-1950s in New York, certain artists began to displace the emphasis in painting onto the act of painting itself. Painting became the document of an anxious activity rather than a visual composition. Among them was Jackson Pollock. Influenced by André Breton's Surrealist theories and experiments in unconscious creation, Pollock produced his consequential *drip* paintings where, instead of touching the brush to the canvas, he placed the canvas on the floor, stood over and on it, and flung/dripped paint onto it, often by making large sweeping patterns. Pollock, via Cubist and Surrealist theories, integrated the tautness of the ground (the flat picture-plane) with illusionistic figural-depth, thereby constructing a tightness of picture-plane (through the extent to which the immediacy of the field is foregrounded) that is at one and the same time non-illusionistic, while simultaneously tending to disintegrate into random noise. The painterly consequences, with their incredibly rich eidetic depth, are suggestive of further abstract noisy spaces and states of mind. This proposal of an oppositional counter-tradition to geometrical perspective places Pollock in a key position to bolster noisy consciousness. The resulting radicalization, as regards their distribution of visual incident into the optical field, *manifests an omni-perspectivalism which is exemplary of the ambient omni-directional aspects of noise.*

As evidence of this trend's beginnings, in 1943 Pollock painted the engulfing *Mural, 1943* for Peggy Guggenheim (1898-1979) in the dimensions of 2.47 by 6.05 meters (about 8 by 19.8 feet). Pollock had been inspired to paint grand-scale works by the murals of the Mexican muralist José Orozco (1883-1949), for example his 1939 copula mural *Mankind* at the Hospital Cabana at Guadalajara, and by the Mexican muralists Diego Rivera (1886-1957) and Alfaro Siqueiros (1896-1974). With the Peggy Guggenheim commission, he transformed the canvas into a whole wall instead of a small object of contemplation which is visually and physically dominated by the viewer. *Mural, 1943* set the precedent for the scale of Pollock's celebrated all-over drip-paintings (with their even distribution of compositional interest across an entire large surface) and it also forced the artist onto the floor for the first time, like the American Indian Navajo sand painters of Arizona and New Mexico whom he admired. Pollock had seen demonstrations of American Indian sand-painting (the making of designs and pictures in sand, made with ceremonial connotations) in 1941 at the Natural History Museum in New York and clearly the ideas behind Navajo sand-painting (its magical and healing aims) influenced Pollock's working method and objectives. By putting the canvas on the floor, Pollock said he could see the painting from all angles, *even from inside it.*

By late-Spring 1947, with the war over and the rebuilding of Europe in process, Pollock began a new series of paintings for his new gallerist Betty Parsons (1900-1982) (as Peggy Guggenheim was moving back to Europe) which was to open January 5th, 1948, in his new (much expanded) atelier, a renovated barn in The Springs, Long Island. As Landau reports in *Jackson Pollock*, we will never know precisely what initiated the radical breakthrough in procedure that occurred between Pollock's last exhibition at Guggenheim's gallery *Art of This Century* and the show he was preparing for Betty Parson's gallery. There are only two documents to help us towards an understanding of why, just then, Pollock strides into what I can justifiably call his *immersive noise* period. One is an application that Pollock prepared in October 1947 as part of a bid for a John Simon Guggenheim Foundation grant (which he was not awarded). As Landau suggests, Pollock's application formulation seems to be inspired by Clement Greenberg's (1909-1994) review in the *Nation*'s art column of February 1, 1947 of Pollock's previous show at *Art of This Century*, in which Greenberg wrote: "Pollock points a way beyond the easel, beyond the mobile, framed picture, to the mural". In accord with this idea, Pollock wrote in his Guggenheim Fellowship statement; "I intend to paint large movable pictures which function between the easel and the mural..". and "I believe the easel picture to be a dying form, and the tendency of modern feeling is towards the wall picture or mural". Pollock went on to further articulate this artistic intention in *Possibilities* (Winter 1947-8), a magazine edited by the artist Robert Motherwell (1915-1991) and the critic Harold Rosenberg (1906-1978). In Pollock's artist's statement entitled *My Painting*, he wrote how he preferred to work on the floor for "on the floor I am more at ease. I feel nearer, more a part of the painting, since this way I can walk around it, work from the four sides and literally be *in* the painting. This is akin to the method of Indian sand painters of the West".

Following the Peggy Guggenheim commission, Pollock desired to paint larger and larger surfaces (almost total environments), especially during the years 1947 and 1948 when he began preparing himself to break with the tradition of European easel painting. Pollock's ensuing appeal for mural commissions increased and in a 1949 letter to his dealer, Betty Parsons, he wrote, "I want to mention that I am going to try to get some mural commissions through an agent. I feel it is important for me to broaden my possibilities in this line of development". The same year Pollock told an interviewer, "The direction that painting seems to be taking is away from the easel, into some sort of wall painting. Some of my canvases are an impractical size ... 9 by 18 feet! But I enjoy working big and whenever I have a chance I do it whether it's practical or not". With the increased size of his canvases, Pollock started to work on them unstretched and to paint them horizontally, laying them flat on the floor at his feet. Then he felt *in the painting*. By working directly on the floor, Pollock was able not only to use gravity to facilitate his radical method of paint application, he was also able to walk around and on the composition, reaching into every part by literally stepping into them.

The viewer of one of Pollock's vast paintings is enticed to recreate mentally (and viscerally) the introscopic noise space Pollock seemed to call upon in the creation of his paintings. Lee Krasner (1908-1984), Pollock's companion and peer, spoke of this space as Pollock's intuitive "pursuit of *unframed space*" which Pollock sought in order to create a spatial continuousness that no longer distinguished between the pictorial space and the area in which the viewer stood. As such, Pollock's imposing paintings demand that the observer relinquish intellectual control (as the beholder is now torn free of unyielding Renaissance perspective) and dive into the energetic noise (through the eye being drawn into the excessive aspect of the painting) and therein dissolve into the dazzling chaos of the individual lines which are also, at the same time, creating a uniformly structured whole-field.

I am reminded here of Wassily Kandinsky (who was a major influence on Pollock) and his book *Text Arista* (that Pollock owned). In it, Kandinsky writes about learning to look at a picture not only from the outside, but to enter it and to move around in it. With Pollock's all-over, syncretistic noise composition, this hypothetical entrance is facilitated as there is no point of reference, no orientation, no parts to its whole[264] (as in Lascaux's Apse). In contrast with the devices of European Renaissance perspective, Pollock sought to draw the viewer into the canvas, not by establishing a distant vanishing-point, but by conceptually eliminating the frame so as to permit the eye to follow the curvilinear patterns beyond the canvas and into the implied surrounding noise space without being stopped by the edges. Here the intention was to create paintings *without beginning or end*, as the vastly increased size of the canvas and its elimination of the traditional frame produced an effect suggesting the ideal immersive 360° noise bubble.[265] This is the unrestrained space of ideal noise, a space where a harmony with the irrationality of noise is encountered. For this, Pollock's work from the late-1940s is an art historical noise watershed.

Relevant also to these concerns are the semi-pejorative statements made by Aldous Huxley (1894-1963) (author of the famous account of a psychedelic encounter with a Belle Portugal rose under the guidance of the Canadian psychiatrist and LSD researcher Humphry Osmond (1917-2004)) concerning Pollock's painting, *Cathedral*, from a 1948 Parsons exhibition. Huxley made these remarks as a participant in the *Roundtable on Modern Art*, a panel discussion held at the Museum of Modern Art from which excerpts were reproduced in *Life* magazine's issue of October 11, 1948. In it, Huxley points out *Cathedral's* lack of focus due to its all-over compositional approach, saying "It raises the question of why it stops when it does. The artist could go on forever. (Laughter) I don't know. It seems like a panel for a wallpaper which is repeated indefinitely around the wall".

Taking this "wallpaper (...) repeated indefinitely around the wall" aspect seriously, the architect Peter Blake (1920-2006), in planning the architectural strategy for what was proposed as the Jackson Pollock Museum, had the idea (with Pollock) to extend the paintings indefinitely around the space. In an article concerned with the project named "Unframed Space: A Museum for Jackson Pollock's Paintings" in *Interiors* magazine, Arthur Drexler (1925-1987) wrote that Pollock's paintings "seem as though they might very well be extended indefinitely, and it is precisely this quality that has been emphasized in the central unit of the plan". About the continuous rhythms of Pollock's paintings, Drexler goes on to describe how, in the model of the museum, "a painting 17 feet long constitutes an entire wall. It is terminated on both ends not by a frame or a solid partition, but by mirrors. The painting is thus extended into miles of reflected space, and leaves no doubt in the observer's mind as to this particular aspect of Pollock's work". In another immersive noise application of Pollock's implied infinity, the artist/architect Tony Smith (1912-1980) designed a hexagonal Catholic church which was to be decorated by stained glass windows executed by Pollock, though the project never went beyond the formulation stage.

It is pleasing to recall that Pollock had written in his Guggenheim Fellowship statement that "I believe the time is not yet ripe for a *full* transition from easel to mural. The pictures I contemplate painting would constitute a halfway state, and an attempt to point out the direction of the future, without arriving there completely". This "direction of the future" was indeed picked up by the generation of artists that ensued Pollock.

The written testimony concerning Pollock's influence derives mainly from the two-part series run in *Art News* in 1967 entitled "Jackson Pollock: An Artists' Symposium" (which included statements by Allan Kaprow (1927-2006), Alfred Otto Wofgang Schulze Wols (1913-1951) and Claes Oldenburg) and Allan Kaprow's "Legacy of Jackson Pollock", also published by *Art News* in 1958, the year that saw Kaprow's first informal *Happening*.

For the artists of the next generation, the generation of the 1960s, Pollock generally represented a liberation of the artwork from traditional means and the inclusion of the artist's life and actions into the work, which lead to other implicit noise conclusions, that is, a freedom from confining structures and the inclusion of movement, gesture, and bodily motion into the realm of visual art. As Kaprow saw it, Pollock "destroyed painting" and freed the painter from working solely in two dimensions. Instead of a 'painter,' one became an 'artist'—capable of working in all and any media.

Noise Event Happening

In his seminal *A Primer of Happenings and Space/Time Art*, Al Hansen states that the idea of the Happening is that of "the artwork enclosing the observer, of art that overlaps and interpenetrates different art forms [...] these performances engulf the spectator: the environment is a work of art that the observer goes into and walks around in and in some cases actually participates in".[266] Generally speaking then, Happenings bombard the participant with an excess of sensations which the viewer has to order in his or her mind to give the overall quality of the continuous commotion (noisily structured like a Cubist assemblage) cohesion. But also Happenings emphasized extemporaneous and migratory elements while manipulating performers, props and audience in ways designed to break down barriers between performance and audience. A Happening was neither an art exhibit nor a theatrical event but an immersive noise site for experimentation in perception.

The prime source of the Happening's central noise concept is that of *collage*, the juxtaposition of unrelated real-life elements in relationships contrived by the artist—that innovation by which Synthetic Cubism had ravaged the Renaissance window-in-the-wall conception of pictorial space. Most often, Happenings placed art inside an *ideal banal sphere* which was imagined less separated from everyday experience, thus challenging the previously established elite hierarchy of values. Towards this end, Happenings were sited in parking-lots, factories or on the street, and involved materials with no fine-art associations. By its emphasis on transient effects and materials, Happenings challenged notions of the permanence of art and the permanence of aesthetic values, hence the Happening became one of the most visible forms of artistic expression of the revolutionary aspect of the 1960s.[267]

Allan Kaprow, in the aforementioned *Art News* article "Jackson Pollock: An Artists' Symposium", explained Pollock's role as progenitor of the Happening thus: "When his all-over canvases were shown at Betty Parsons's gallery around 1950, with four windowless walls nearly covered, the effect was that of an *overwhelming environment*, the paintings' skin rising towards the middle of the room, drenching and assaulting the visitor in waves of attacking and retreating pulsations. [...]. The expanding scale of Pollock's work, their reiterative configurations prompting the marvellous thought that they could go on forever in any direction including out, soon made the gallery as useless as the canvas, and choices of wider and wider fields of environmental reference followed. In process, the Happening was developed". However, in *Pollock Painting: The Photographs of Hans Namuth*, Barbara Rose proposes that it was the publication of Hans Namuth's photographs and his film of Pollock painting that are responsible for the development of Happenings (as well as anti-form, distributional, conceptual, performance, and body art). But whatever the specific rationale, the implications of Pollock's work were vast, exerting even a persuasive impact on avant-garde dance, as it has been often noted that the dance choreography of Merce Cunningham is closely related to Pollock's painting. Cunningham essentially fused noise ideas extrapolated from Pollock with those of Marcel Duchamp, as understood and practised by his collaborating composer, John Cage. This tendency in dance was explored in the early-1960s by Jill Johnston, Yvonne Rainer and Ann Halprin.

The feeling/concept of "space in which anything can happen" seems to summarize Pollock's general significance to the artists he inspired. As an example, Claes Oldenburg, self-described as a "Post-Pollock painter" in the 1967 *Art News* article, conceived of himself as standing on the canvas that became his surroundings, and which stretched as far as he could see or hear. This suggests the idea of a *new theme of distribution* where the city's many signs are no longer depicted but included in the work, hence immersing the artist (and viewer's attention) in a new (Pop) art based on reproduction.

Certainly by the mid-1950s, media (print, radio, and television) influenced almost everything everywhere in post-industrialized countries, including members of the Gutai Group of Osaka Japan, an art-theater group made up of painters (including Akira Kanayama, Sadamasa Motonaga, Shuso Mukai, Saburo Mirakami, Atsuko Tanaka, Shozo Shinamoto and Kazua Shiraga). They had seen photographs of the theatrical French theoretician and action painter Georges Mathieu in *Life Magazine* decked out in an elaborate costume painting before television cameras at the Sarah Bernhardt Théâtre in Paris in 1956. These photos inspired their own live painting performances in which they threw balls of paint at the canvas or, in another instance, where an artist ran and leaped through a series of sequential canvases.

Moreover, Kazua Shiraga, a member of the Gutai Group, adapted and exaggerated Pollock's painting techniques and Mathieu's theatrical presentation of the painting-action with his *Making a Work with His Own Body* (1955), where the artist wallowed in the paint medium with his entire body, and again at the Festival di Osaka's *Painting Performance* (1959). Hanging from a rope, Shiraga threw himself, in a kind of overwrought psychic automatism, on the canvas and spread lumps of color with his feet around while swinging on the rope. Here the artist is literally inside the painting in a way which we will see again with the Vienna Actionists. Shiraga's work predates and reminds us of Yves Klein's well known 1960 *Symphonie Monotone* (Monotone Symphony) painting/music performance at the Galerie International d'Art Contemporain in Paris, especially when we consider it from the immersive position of the three female models themselves (one of whom I met and discussed this with).[268]

Another earlier moment leading to the program of the Happening can be traced to an evening in 1952 organized by John Cage at Black Mountain College. For the performance, an audience was seated in four inward-facing blocks as Cage delivered a lecture, punctuated by silences, from the top of a ladder. Poet Charles Olsen (1910-1970) and others read poems from another ladder while David Tudor played a piano and Robert Rauschenberg played a wind-up gramophone. Through this rich conflicting event, Merce Cunningham and other dancers moved about through the space where some of Rauschenberg's early white-on-white *White Paintings* were suspended as a sort of false ceiling overhead.[269]

In the mid-1980s, I obtained one of Rauschenberg's black and white silkscreens from his *Current and Surface Series* (1970). Living with this work, I discovered the best of Rauschenberg's noise work I think—work that contains rhizomatic layered image sequences where the viewer interprets the progression of images as though reading a ruined communication system arranged in multiple, simultaneous combinations. Rauschenberg here dissolves away the paradigmatic model of media as communications and replaces it with one of failed pageant that leads to both a collapse of meaning and the destruction of distinctions between media and myth. In Rauschenberg's media noise society, I saw through the numerous saturating media messages, so that information and meaning imploded into pure effect, without content or meaning. In fact, here content becomes decorative and ornate.

This noisy rhizomatic Rauschenberg demands a different kind of looking—akin to the aggregated viewpoints of Cubism compared by John Cage to watching "many television sets working simultaneously all tuned in differently".[270] In my piece from his *Current and Surface* Series, there is no obvious hierarchy of images to scan. The trajectory of visual exploration for it is of our own choosing—a dysfunctional situation that no longer communicates purposeful messages but rather proposes noise pattern. Here we have a rhizomatic visual pleasure, where everything equally connects to everything else and so replaces visual purpose. This is a noise art that demands of society an active visualizing participation in private interpretations—and thus is a legitimate metaphor for contemporary art as a form of simulation-shattering engagement. It functions by overloading representation to a point where it becomes non-representational noise.[271]

John Cage's strict musical development does not concern us here, except for his idea—derived from the work of Edgard Varése and the Zen philosophy of Daisety Taitaro Suzuki (1870-1966)—of treating all forms of noise as sound to be used by the composer, together with the corollary that silence is just as important. This led him ultimately to the *nec plus ultra* of noise music, 4' 33", perhaps the musical equivalent of the white-on-white canvases of Kazimir Malevich (1878-1935). 4' 33" consists of "silence" performed for this duration of time. In 4' 3", the fortuitous immersive noises in the room, usually unnoticed, and the hearers' own thoughts, become the content of the piece.

In terms of one's own thoughts becoming the content of a piece, we must note that in 1954, Dr. John C. Lilly, a pioneer in brain and behavioral research studies, began experimenting with the concept of restricting the amount of external sensory stimuli to the brain in a kind of anti-noise research project. When Lilly built his first isolation environment (what came to be known as isolation float tanks) he was determined to prove that the brain, without environmental input, would simply go to sleep. Using his own being for experiments, he learned the opposite is true. By removing all visual, acoustical, tactile and temperature stimuli, Lilly found that the brain continues to function independently and at an even higher level than normal. I have had this experience myself, so I know it to be true.

But the artist most identified with the external noise Happening and perhaps its chief exponent is John Cage's composition class student, Allan Kaprow. Kaprow began as a painter and his paintings moved from Abstract Expressionism into increasingly complex action-collage assemblages, like *Pentiy Arcade* (1956), *Wall* (1959) and *Kiosk* (1959), which were developed following his interest in the work of Jackson Pollock. The action-collages became bigger and projected further and further from the walls and into the room and included more and more audible elements. A person entering an action-collaged-noise space would become lost in an excessive labyrinthine atmosphere. Kaprow eventually thought how much better it would be if a visitor could just go out of doors. Thus in Kaprow's form of the Happening, ordinary people, ordinary time, and the everyday spaces of street and supermarket noise, were frequently merged.

Kaprow became a professor of art history, and this academic side of his activities made him a fluent and perceptive theorist, enabling him to elucidate how Happenings evolved from the action-collage environment idea of opening up the mind and the eye to the world of the street. In his book, *Assemblage, Environments and Happenings*, Kaprow explains the fusion of the concepts behind Pollock's gestural paintings and the junk-assemblage sculpture movement as culminated in this aspect of the Happening. Kaprow also wrote in his famous 1958 essay "The Legacy of Jackson Pollock" that "Pollock, as I see him, left us at the point where we must become preoccupied with and even dazzled by the space and objects of everyday life".

By 1957, Kaprow's work became exclusively environmental involving lights, odors, electronic sounds and unusual materials. His environment at the Hansa Gallery in 1958 contained no art objects as such, but initiated a conception that *art was experienced as a surrounding* rather than a picture or sculpture to be looked at, a surrounding that engaged the visitor with things to do. However, Kaprow's first mature noise Happening, which involved Dada provocation, assemblage, and action painting, was *18 Happenings in 6 Parts* that took place at the Reuban Gallery in October 1959 in New York City. It was tightly scripted and drilled.[272] After *18 Happenings*, Kaprow did a number of similar pieces, including *Coca-Cola, Shirley Cannonball?* and *A Spring Happening*, the latter taking place in the new downstairs premises of the Reuben Gallery in March 1961. After waiting in a curtained-off lobby, the audience was shown into a dark tunnel, made of wood and hardboard painted black, and with a slit running along both sides at eye-level where they remained for the duration of the piece.

For the next seven years, Kaprow expanded the potential of the *environment* in a gallery setting, but gradually the showroom space was abandoned for more informal and natural settings such as vacant breweries, open fields, and woods. By 1969, Kaprow's work had evolved so distinctly into new phases that he gave up the designation *Happening* and adopted Michael Kirby's (1931-1997) term *Activity*.

By the early-1960s, artists' noise actions performed in front of audiences and or cameras became more and more familiar as the Happening movement gained momentum. Two types of Happenings emerged: one involving a more or less static audience, and the other a walk-around environment (like Kaprow's *Words,* which was installed at the Smolin Gallery in 1962). *Words* was an arrangement of audience-participation devices, rolls of words to move, words on cards hung on strings, words to pin up and rubber-stamps to make phrases with. *Garage,An Apple Shrine* and *Yard* (which filled the Martha Jackson Gallery with car tires) also utilized such an approach.

In Paris during the late-1950s, a Marcel Duchamp-inspired renewed interest in Dada noise gave rise to various actions, dé-collages, and performances by artists such as Robert Filliou (1926-1987) and Jean-Jacques Lebel, the most active member of the group of younger artists to emerge from the Nouveaux Réalistes precepts. As an example of the first approach to the Happening, in an early 1960 Happening *Funeral Ceremony of the Anti-Process* conducted in Venice, Italy, Lebel invited the audience to attend a ceremony in formal dress. In a decorated room within a grand residence, a draped 'cadaver' rested on a plinth which was then ritually stabbed by an 'executioner' while a 'service' was read consisting of extracts from the previously mentioned French decadent writer Joris-Karl Huysmans and the Marquis de Sade (1740-1814). Then, pall-bearers carried the coffin out into a gondola and the 'body,' which was in fact a mechanical noise music sculpture by Jean Tinguely (1925-1991), was ceremonially slid into the canal.

Conspicuous, too, in this regard was the fascinating technologically-aided presentation of Mark Boyle, the performance *Son et Lumière for Body Fluids* (1966), where he presented a heterosexual couple making love with their encephalograms projected and enlarged on a screen above them.[273] Boyle went on to create light-shows for the psychedelic rock group, Soft Machine, and was involved in an early British experimental night-club called UFO.

Another important early noise Happening artist is Carolee Schneemann, particularly with her highly immersive (for the participants) and spectacular Happening bacchanal called *Meat Joy,* performed at the Judson Memorial Church in New York City in 1964 and in various locations in Europe (including the *Festival de la libre expression* at the American Center in Paris in 1964).

The Dionysian mystical impact of Schneemann's *Meat Joy* was heightened by the sexual implications of voluptuous, scantily clad people wallowing provocatively in paint and meat, somewhat beyond the truism that all sexual activity is about the mixing of gametes. Schneemann's environmental noise performance (performed in what she characterized as a "sensory arena"), *Illinois Central*, utilized a 360° visual environment contrived with film and slides that shifted over time.[274] Schneemann's grotto-like niche entitled *Up To and Including Her Limits (Trackings)*, which she built for herself at the Basel Art Fair in 1976 also impresses, as the work addresses noise as liberation from confine. So, too, does David Tudor's *Rainforest IV* of 1973 (realized by Composers Inside Electronics), as the viewer is an integral part of the work. *Rainforest* extended the implications of Erik Satie's ambient *Furniture Music* of 1920. Like Satie (whom Tudor admired) *Rainforest IV* overturned the traditional view that music is performed at a specific time in a proscenium space in which the performers and audience are separate.

This relates to Yoko Ono's *Cut Piece*, first performed in Tokyo in 1964, in which she invited the audience to cut her clothes off, and deserves reference in terms of an artist putting herself in a visceral environment with a high-resonance of associative connotations. The same holds true for Yayoi Kusama, whose theoretical polemic concerning the *distributed, scattered, multiplied, and obliterated self* is best established, as she herself states, as an "obliteration of everything (including myself and others)" into a beguiling and excessive artifice which gives birth to an *opalescent non-existence*.[275] In explanation of her installation work, Yayoi Kusama said, "One day, I was looking at a table cloth covered in red flowers, which was spread out on the table. Then I looked up toward the ceiling. There, on the windows and even on the pillars, I could see the same red flowers. They were all over the place in the room, my body, and entire universe. I finally came to a self-obliteration and returned to be restored to the infinity of eternal time and the absoluteness of space".[276] Paradoxically, Kusama tried to achieve an expression of this idea of the obliterated self by exposing herself (and others) fully nude and painted with polka-dots in various Happenings at high-profile New York City locations. Kusama staged several public demonstrations of painted polka-dotted nakedness entitled *Anatomic Explosion*, most notably on Wall Street, in the sculpture garden at the Museum of Modern Art, and in 1968 at the statue of Alice in Wonderland in Central Park.

Certainly, we must also briefly recall the Happenings of Jim Dine, most notably *Car Crash* (1960) and those of Wolf Vostell, for example his *You* (1964). Important, too, were the Happenings of Al Hansen, Dick Higgins (1938-1998), Claes Oldenburg, Red Grooms, Robert Whitman, Meredith Monk, Jeff Nuttall, John Latham and later the group improvisations of the Movement Collective.

But the Abstract Expressionist ancestry of noise as aesthetic experience is not confined to post-Pollock aesthetics. In this respect it is significant to note that a motion-picture immersive environment called *Impressions of Speed* appeared at the 1958 Brussels World Fair which seated 25 people at a time in the cab of a simulated railroad-engine. On view via wide screens was a color landscape in the front and on the two sides as well; a *continuous, all-encompassing image* projected on the simulated windows in an attempt to duplicate the total impression of actual peripheral noise experience.

To review the history of painting in relation to aspects of noise, I would be remiss to neglect to mention Post-Abstract Expressionist applications as practised by Robert Ryman. The 'pure' opticality of the color white embraced by Ryman is a prime example of subtle noise, as in his white paintings he addresses the problems arising from the tension created from the opposition between surface materiality and opticality in relationship to the edge of the painting and its relationship to the wall on which it is hung. This ambiguity of the painting's boundary in relation to the wall that contains it draws attention into an expanded subtle noise field which we will see will come to define the immersive art of the 1960s and 70s. Ryman does this by extending the optical white shimmering-field of color/light out from the painting onto the white gallery walls which present it. Now it is really the wall that provides the painter with his ground which Ryman himself clarifies when he writes "the wall plane is actually part of the painting and it extends out three or four feet...". Hence Ryman presents his paintings as part of the white cube that has come to represent modernist ideals of purity and neutrality. The whiteness of the paintings require the whiteness of the walls, as the white-painted optical field spills out over the confining edge of the painting to fill, theoretically, the entire wall and room, thus texture, surface-plane, color and wall are unified. As Ryman himself says: "The wall becomes very much a part of the work", and so by blurring the difference between painting and wall, Ryman extends our consciousness of painting into an expanded, immersive, subtle, visual noise environment. Of course, this liberation of color from form in the service of filling a room can also be seen in the neon-tube installations of Dan Flavin, where color spills out over the walls of the gallery in which the piece is installed, expanding its presence dramatically and soaking the visitor to the space in its soft, vibrating light.

So in the Viennese Actionist disposition, to move away from Abstract Expressionist action painting in the 1960s and towards the performance-oriented tendency of *Actionism*, was for the Viennese Actionists, very much in stride with the significant art of their era, impelled, as they were, by a Herculean sense of noisy idealism based on a felt necessity for emancipation from what they saw as the repressive constraints of church and state power. Consequently, their Actions were intentionally and noisily inciting: deliberately exhibitionist, abhorrent, sexist and/or sacrilegious.

In a sense, Actionism can be seen in retrospect as a logical extenuation of the heroic male individuality of the Abstract Expressionist generation and their idealistic attempt to create a new post-war world based on an intimate subjectivity in pursuit of societal freedoms by turning their back on ideological traditions and engaging in the supposed non-ideological material world of the immediate. Though this seems an overly naïve belief to us now, it did provide the idealistic engine to what became a body of incredible noise work. In the early 1960s, the Actionists Günter Brus, Otto Mühl, Alfons Schilling and Rudolf Schwarzkogler began sensing their late connection with the Abstract Expressionist movement when already the arbitrary nature of personal subjective expression was beginning to become apparent in the repetitions of what became the Abstract Expressionist gestural formula. By the time the Actionists engaged in it, what was originally hailed as a new common language, gestural abstraction, began to degenerate into a self-indulgent, dipsomaniac activity in the hands of the more recent Abstract Expressionist neophytes. To their credit, the Actionist artists began to see that the total reliance on Abstract Expressionism's subjective feeling of personal assertion (which surprisingly began to look ever more and more similar) meant that Abstract Expressionism's message of immediacy and physicality was arbitrary. To counterbalance this, the Actionists, in a peculiarly comparative manner to the Pop and especially the Fluxus artists, aimed to produce art closer to real life and to re-mix aspects of reality into their art.[277] Thus they moved away from Abstract Expressionist ideology and eventually towards a greater "objectivity" of real life, which in turn led to the urge to challenge the power structures of church and state. Therefore, the Actionists moved art away from represented conflict (as recorded on the Abstract Expressionist canvas) and towards political conflicts and social associations in life between people.

In the Viennese studio, Günter Brus had been drawn towards Abstract Expressionist type *informel* painting and, following Pollock's lead, began identifying himself as working from *inside of nature*. Early on, he exemplified this ideology in his *Labyrinth Paintings* which he executed through the means of disorientation in immersive space. In the autumn of 1960, Brus almost entirely cleaned out his 2.5 by 6 meter (roughly 8.2 by 19.6 feet) painting studio and placed white-painted paper over all the available walls and began making use of the entire room (from floor to ceiling) in the unfettered splattering application of black paint, utilizing all three of the available surfaces simultaneously in an attempt to fracture the domination of the compositional mid-point and to penetrate into a much fuller sensation of immersive space. By doing so, Brus developed the ideal of the all-pervasive sphere in which the artist would be enclosed and in which the artist would then paint thoroughly in three-dimensions, using both feet and both hands.

Günter Brus's close painter friend at the time was Alfons Schilling, an artist who went on to utilize a mechanical machine in the creation of his paintings and who still later developed a brilliant series of consequential FOV modifying viewer head-pieces.

As documentation of the ideals under pursuit in the Actionist circle, Schilling left us some interesting extracts from his notebook from early-1961, which also shed light on the issue of immersive thinking in post-Abstract Expressionist painting. In them he wrote, "I can only feel infinity if I break out and reach beyond the closed composition and the frame. [...]. One must be able to enter my pictures from all sides and be able to leave it from all sides; the picture then continues like a tone that has been struck. [...]. The possibility of a limitless, never-ending painting can only be represented by means of a section. How can I possibly perceive 'infinity' in a picture, as long as the possibility of seeing pictures as something complete in themselves, is still not removed. Every barrier must be removed from one's vision (even if it is only the edge of the picture). A picture must offer no opportunity of beginning or ending anywhere. [...].

Getting inside, being inside, and having achieved unity I experience everything in a state of transformation".

In 1963, Günter Brus received 5000 schillings from the Institut zur Förderung der Künste to assist him in the creation of a series of large-scale paintings. To do these large paintings, he stretched string backwards and forwards across empty gallery rooms and hung molino (a cheap substitute for canvas) and paper so that they reached the floor in order to create a labyrinth which would help prevent him from preconceiving a compositional idea too quickly. He then painted all the surfaces as if it were one large painting that completely surrounded him. Few people saw the painted labyrinth, however.

Subsequently, in the autumn of 1964, Brus carried out his first real Action titled *Ana* which took place in Otto Mühl's studio, a fellow artist and friend. In preparation for *Ana*, Brus painted Mühl's studio and several objects in the room (typical of a Viennese bourgeois apartment) a stark white. In effect, he began his Action with the classic white canvas, now extended out into the third-dimension. Hence he begins in an enveloping, immersive, unified, total-space. On starting the work, he emphasized this enveloping further by rolling across the floor of the room with his body completely wrapped in pieces of white cloth. The pieces of cloth unwound as a result of the motion and he remained motionless for a long period of time. Then Brus began to stream black paint over the white objects and over his wife who also participated (passively) in the action, with the aim of making a living painting. He then burst into a bout of painting and besmeared the walls until exhausted.

After *Ana*, Brus decides to produce the action called *Self-Painting* in which his own physique was to serve as a painting surface with the intent of binding himself into the picture-plane in order to "become one with the picture" and to thereby "disappear into the picture". words which remind us of Yayoi Kusama's avowed ideal of doing likewise. As mentioned, Kusama has described the emergence of this perception/ideal by recounting a moment when she was watching a red pattern of a tablecloth coat everything around her and then swallow her up this way; "When I looked up, I saw the same pattern covering the ceiling, the windows, and the walls, and finally all over the room, my body and the universe. I felt as if I had begun to self-obliterate, to revolve in the infinity of endless time and absoluteness of space, and be reduced to nothingness".[278] With apparently similar aims, Brus designed *Self-Painting* as a soundless action separated into three separate tableaux in which Brus placed together different parts of his painted white body with disparate objects that were also painted milky white. A jet black streak is painted vertically over Brus's face by himself and along his forearm as if his body had been ripped open by one of Barnett Newman's majestic zips.

In January 1965, Brus went on to perform painting actions *Silver*, *Self-Painting II* and *Self-Mutilation* for a film-maker and photographer and, on July 6th, 1965, he performed a *Self-Painting* at the Galerie Junge Generation in Vienna. The day before, on July 5th, he painted himself as in the *Self-Painting* Actions and proceeded to stride across Vienna as a living painting, but was stopped and arrested by a policeman for causing a civic disturbance.

Brus's peer Otto Mühl, too, was coming from the process-oriented *matière* side of *Art Informel* and the related *assemblage* movement of the Nouveaux Réalistes (for example Arman's *accumulations* of everyday rubbish) and the American *junk sculpture* movement. The term *assemblage* was coined in 1953 by Jean Dubuffet (1901-1983) to refer to works that supposedly went beyond the collage of Synthetic Cubism. In junk/assemblage sculpture of the late-1950s (and with Robert Rauschenberg's *combines*) art further challenged the boundary between everyday objects and High Art and the entire world opened up and became the raw material for the creation of art.

Mühl had met Brus in early December of 1960 at the famed Gesamtkunstwerk-oriented Sezession building and they shortly thereafter became engaged in an artistic discourse which eventually indelibly shaped both men's work. Otto Mühl wrote in "Weg aus dem Sumpf" in 1977 of Brus that "Brus painted in psychomotoric expressionist style, a wild criss-cross of lines hurled onto the paper. The paint sometimes exploded like a bomb when it hit the picture. That was total creative excess. I understood right away and was full of enthusiasm. The pictures were often 5 meters long by 3 meters high. The whole room was covered with splatters of paint, on the floor there was a centimeter thick layer of paint ooze that had dried up". In 1961, Mühl gave up traditionally-scaled easel painting and began a series of Actions in which he poured paint and pigment onto paper and then wallowed in it, bringing structure to the pools of color. In a letter to his friend Erika Stocker dated January 8th, 1961,

Mühl writes, "I have, so to speak, produced my first *tachiste* picture. To do it I have developed an original technique. I painted it by laying it on the floor. It doesn't work on the easel anymore". And on March 23rd, 1961 he writes her again saying, "I wallowed in paint [...]. I slid from one end to the other, turning over once. In the process I worked the surface with my hands... ". Taking this sense of wallowing further, in May of 1961, Mühl created a full room installation in his studio called *The Overcoming of the Easel Picture by the Representation of its Destruction Process* by nailing or tying together his paintings into a unified Gesamtkunstwerk. Following this installation, Mühl moved increasingly towards creating three-dimensional assemblages and from there deeper into performance Actions.

Disappearance of the Noisy Art Object into the System

In retrospect, the shift in art in the 1960s and 1970s towards an open, more immersively inviting noise dominion of self-attentiveness, with its emphasis on recontextualization and release from the framing apparatus, can be seen as an anticipation of and desire for omni-directional ambient noise consciousness, with its ideal 360° bubble-like vista. This optic, which is located radiating out in all virtual directions at once, can be seen as a further extenuation of the expanded field that cybernetic-influenced art instigated.[279]

The disappearance of the *objet d'art* in roughly 1965 marked the emergence indicative of post-modernist immersive noise experimentation, which was postulated on the assumption that the art experience needed broadening. Frank Popper's seminal book *Art-Action and Participation* is an important reference to this development, as is Jack Burnham's book *Beyond Modern Sculpture*. Burnham arrived at the conclusion that cybernetic sculpture, or rather the cybernetically-informed sculptor, is not simply adopting new materials and new standards of fabrication, but evolving a new aesthetic, now synchronized with technical ideals. Cybernetics had demonstrated that the configuration of a system is an index of the performance that may be expected from it, hence cybernetics' extremely circular-state yields an extended aesthetic noise consciousness on the basis of connected self-attentiveness, and it is within this elastic self-attentive aesthetic framework where we will expect to find new noise attitudes emerging in art.

The recontextualization of the *objet d'art* into the global envelopment of the noise environment (where the viewer is pulled away from the constraining aperture of the picture frame and more and more from the gallery frame) is indicative of the immersive qualities of the era under investigation here. This radical deframing opened up the viewing cone of the 1950s' post-cubist/post-war painting space towards a more thorough literalization of the imagined (or implied) non-partial field of universal noise surroundings/conceptualizations of abstract space. Here, framed areas of noise may not be singled out and made to represent the totality of range.

This noise immersive space, where partial framed and arranged views may not be cut out of the total surround, finds a very real literalization in the open field of art in the 1960s and 1970s, and the broad holonogic gaze which it provokes is a huge step in the direction of escaping the limits of narrow representation in the interests of hyper-noise consciousness. From this point on, only a technique that fully undermines the proscenium and window-like frame can stand in for the abstract, all-over, intemperate 360° bubble-noise which the frame cuts and excludes. In this drift towards anti-representationalism, noise art begins leaving the orbit of the framing apparatus and of the tunnel vision that fixed a segment of the objective world at one end and the viewer at the other. What had enabled that narrow cone of vision to simulate the entire visual atmospheric field previously, was possible precisely with the enclosure of that framing cone (tangent tunnel). But once that framing cone has dissolved through Kant's indeterminate supersensible, noise's distributed spatiality, expansion, dematerialization, excess and/or any other number of Op, Cybernetic or Conceptualist artistic strategies, that narrow cone of representation is found wanting and a much more encompassing, atmospheric, scopic hyper-noise art is conceivable.

Art in the 1960s' open arena, then, is generally conceived of as a noise cluster of optical vectors which suggest a hyper-total, enveloping, non-vectored space that creates unaccustomed situations and sensations for the enthusiastic in an attempt to shift the political/social vortex away from outdated symbolic allegiances and towards sensate dynamic forces of change. As such, it stands in contrast to the standard histories and doctrines and ideas that were being propagated in the mass media at the time.

Here I will review some pertinent noise examples concerning the semi-disappearance of the *objet d'art* which occurred in the late-1960s that to a great extent set the tone for vanguard art on through to the late-1970s when a revival of painting occurred under the designation of post-modernist appropriation/simulation. Of course, artists continued to paint and sculpt throughout the period of the 1960s and 1970s, and still up to this day, but they do so from a derrière-garde position, as traditionalists. This is so because a change in art occurred in the late-1960s when art typically lost its artisanal materiality (as discrete paintings and sculptures) and became increasingly time-based and ambient as a repercussion (primarily) of the legacy of painterly abstraction in the early and mid-20th century. In terms of the immersive noise inclination, this expansion away from the two-dimensional canvas freed the spectator from stasis and encouraged an active atmosphere of contemplative reception *within the work of art* which was essentially attained through the compliant motion of the immersant in contact with the strategic liberties exacted in the expanded art. Artists increasingly aimed in this era to evoke noise possibilities within the imagination of their audience, to engage their active participation and to release art from its previous obligatory fidelities to the hypothetical and material status quo. Underlying this aim is a miasmatic noise idea which questions linear and hierarchical structures and seeks to replace them with atmospheric loose structures, keyed to a penetrable, reciprocal flow of events. This inclination might be further characterized as the deposit of hyper-noise within the immersant that will manifest at a later date as a personal and private inner art: the creation of future noise artists, in other words.

Much of the disappearance, de-definition and de-materialization of the art object that went beyond Modernism in the search for a total art of noise developed out of the visual spectator's participation that was called for in viewing *Op Art*: a hard-edge geometrical movement that flourished in the early-1960s (largely inspired by various optical experiments of Marcel Duchamp) in the work of Jesus-Rafael Soto, Bridget Riley, the GRAV group, Yayoi Kusama, Yaacov Agam, Pol Bury, Josef Albers, Marian Zazeela and Victor Vasarely, among others. Op Art called attention to the spectator's individual, constructive and changing perceptions, and thus called upon the attitude of the spectator to transfer the creative act increasingly upon him or herself. This ideal, in turn, beckons forth a consideration of the enlargement of the audience's normal participation, both in regard to the spectator's ocular aptitude to instigate variations in the perceived optic, and to his or her capability to produce kinetic and aggregate exchanges on or within the work of art itself.

Indeed, Kinetic Art also played an important part in pioneering the unambiguous use of optical movement and in fashioning links between science, technology and art relating to the notion of the environment. Simply stated, the term *kinetic* means the study of the relationship between moving bodies, hence the term *Kinetic Art* is usually used to describe either three-dimensional mobiles or constructions that move in either foreordained or unplanned ways. With Op Art (which is kinetic in that Op situations employ optical illusion that effect an appearance of motion) and Kinetic Art (both conceptual descendants of the shifting noise perceptions initiated in 20th century painting with Impressionism, Cubism and Futurism), the artwork under consideration is no longer merely a categorical system but increasingly an *invocation to noise perception*. The cognitive noise encounter that a spectator may undergo in an Op situation, perhaps best exemplified by Bridget Riley's projected circular Op environment done for the 1960 exhibition *Situations* in London, was instigated by the certitude that the spectator was obliged to take up consecutive positions in front of the display, in order to detect the series of shifting patterns and lines that were offering themselves to the onlooker from contrary and incompatible angles. Thus the element of personal choice and physical motion by the beholder is emphasized, resulting in a decline in the art object's sequestered, fetishistic standing as an *objet d'art*. This is well exemplified, too, by Jesus-Rafael Soto's process-based walk-through Op environments called *Penetrables,* which incorporated a tactile immersion into visual noise (with occasional sonorous elements) notable for their immersive noise attributes (given their realization on an architectural scale). The work increasingly becomes a co-operative production of the operation between the former *objet d'art* and me, as I am *projecting my selfhood into the noise form* and am thereby enabled to sense the various spatial possibilities the shifting hyper-noise suggests.

Many sensory noise projects, installations and environmental events produced by the Brazilian artist Hélio Oiticica (1937-1980) excellently exemplify this trend. In Oiticica's work, the once established correlations between the spectator, the *objet d'art* and the artist is radically modified.

With Oiticica, the emphasis is no longer on the *objet d'art* created by the artist, and certainly not solely on the personal fancy of the immersant, but on a third dramatizing maneuver similar to what Brion Gysin (1916-1986) and William S. Burroughs call *the third mind*. The third mind is based on Brion Gysin's rediscovery of Tristan Tzara's (1896-1963) Dada cut-up writing method which he encountered while cutting through a newspaper he was using to trim floor mats. Gysin did several experiments with cut-ups while living in Tangiers and shared them with his friend William S. Burroughs. Thereafter Burroughs used cut-ups in his books *Nova Express, The Ticket That Exploded*, and others. Gysin, too, was responsible for the absolutely noise-immersive optical *Dream Machine* that he invented based on the sparkling and flickering of the sun through the trees.

The principle behind the *Dream Machine* is that it generates wave-like patterns which strobe at around 10 Hz, the frequency of the alpha waves sometimes present in the part of the brainstem responsible for determining states of creative consciousness. As one sits (relaxed) in a room filled with the machine-generated flickering light, spectacular hyper-noise visualizations may occur due to the optical twinkle at work.

When I saw the *The Third Mind* exhibition in the fall of 2007 at Le Palais de Tokyo in Paris (curated by Ugo Rondinone) many noise issues arose in my mind. The show contained work from: Ronald Bladen, Lee Bontecou, Martin, Boyce, Joe Brainard, Valentin Carron, Vija Celmins, Bruce Conner, Verne Dawson, Jay Defeo, Trisha Donnelly, Urs Fischer, Bruno Gironcoli, Robert Gober, Nancy Grossman, Hans Josephsohn, Brion Gysin and William Burroughs, Toba Khedoori, Karen Kilimnik, Emma Kunz, Andrew Lord, Sarah Lucas, Hugo Markl, Cady Noland, Laurie Parsons, Jean-Frederic Schnyder, Josh Smith, Paul Thek, Andy Warhol, Rebecca Warren, and Sue Williams.

What is interesting about this disquieting show is to look at how *this* group show differs in its conjoining (or not) from other group shows by pinning it to the collaborative work of Brion Gysin and William S. Burroughs from the early 1960s known as *The Third Mind*. Moreover, can we place this *Third Mind* in the context of wider noise connections and ponder at what point does homage turn into exploitation?

Burroughs and Gysin, known predominantly, as mentioned, for the rediscovery of the Dada master Tristan Tzara's cut-up technique and for co-inventing the flickering *Dream Machine* device, worked together in the early 1960s on a publishing project that used a chance-based cut-up method. A cut-up method consists of cutting up and randomly reassembling various fragments of something to give them a completely new and unexpected meaning: 1+1=3. In the biography of Allen Ginsberg (1926-1997), *Celebrate Myself*, Ginsberg's archivist, Bill Morgan, recounts some of the geneses of Brion Gysin and William S. Burroughs forays into radical Dada cut-up technique and collaboration based on Ginsberg's diary entries.

In the mid-1950's, Gysin pointed out to Burroughs that collage technique has been a regular tool in painting and graphics since half a century. This came as late news to the young Beat writers of that time, so it is perhaps not surprising that Ginsberg's first exposure to Burroughs's use of the cut-up was met with disdain—Ginsberg considered it something along the lines of a parlor trick.[280] Even more, Ginsberg speculated from NYC that Burroughs had lost his mind through lack of sex.[281] As a joke, Ginsberg and Peter Orlovsky cut up some of their own poems and rearranged them and sent them to Burroughs with the note, "Just having a little fun mother".[282] However, Burroughs was so dedicated to the random cut-up method that he often defended his use of the technique. When Ginsberg and Orlovsky arrived in Tangiers in 1961, Burroughs was working on an even more advanced use of the cut-up; he and Ian Sommerville (1940-1976) were cutting and splicing audiotapes and Burroughs was making collages from newspapers and photographs while proclaiming that poetry and words were dead.[283]

Burroughs, however, soon began work on a cut-up novel, the *Soft Machine*, drawing material from his *The Word Hoard*. *The Word Hoard* is a collection of Burroughs's manuscripts written in Tangier, Paris, and London that all together created the mother-load manuscript that served as the basis for much of Burroughs' cut-up writings: *The Soft Machine, Nova Express, The Ticket That Exploded,* (together referred to as *The Nova Trilogy* or *Nova Epic*). Even *Naked Lunch* was taken from sections of *The Word Hoard*. A text was also produced called *Dead Fingers Talk* in 1963, which contains excerpts from *Naked Lunch, The Soft Machine* and *The Ticket That Exploded* combined together to create a new narrative. Also, via Burroughs's artistic collaborations with Brion Gysin and Ian Sommerville, the cut-up technique was combined with images, Gysin's paintings, and sound, via Somerville's tape recorders.

The *Soft Machine* manuscript was soon being assembled and edited by Ian Sommerville and Michael Portman, Burroughs's companions. Sommerville was regularly speaking of building electrical cut-up machines. Shortly thereafter, Burroughs would begin collaborating on a book project with Brion Gysin using the cut-up method, cutting up and reassembling various fragments of sentences and images to give them a new and unexpected meaning. *The Third Mind* is the title of the book they devised together following this method, and they were so overwhelmed by the results that they felt it had been composed by a third person; a third author (mind) made of a synthesis of their two personalities. Ginsberg remained highly skeptical for some time, but following his travels in India came to appreciate the cut-up technique, even while never employing it.

Now for *The Third Mind* show. Many artworks found here advance Rondinone's thesis of the third mind. Of course, foremost is the Brion Gysin and William S. Burroughs collaboration, *The Third Mind*. An entire gallery is devoted to the maquettes for this unpublished book from the collection of the Los Angeles County Museum of Art—and it does not disillusion the fourth mind: that of the viewer/reader. It is a golden hodgepodge feast and serves as the noise underpinnings of the exhibit.

Then there is the glamorous video installation/accumulation of Andy Warhol's (1928-1987) *Screen Tests* from 1964-1966: a group of silent black & white three-minute films in which visitors to the Warhol factory try to sit still. Here we see an interlaced presentation that visually connects the youthful faces of Edie Sedgwick (1943-1971), Susan Sontag (1933-2004), Nico (1938-1988), John Giorno, Jonas Mekas, Gerard Malanga, Jack Smith (1932-1989), Paul Thek (1933-1988), Lou Reed and the distinguished Marcel Duchamp. The presentation is structurally connectivist given its four-directional presentation as a low laying sculpture. It is incredibly enjoyable. Plus the room is ringed with black haunting photograms called *Angels* by the fascinating Bruce Conner (1933-2008) from 1973-75.

In terms of a more traditional, synthetic, associational, curatorial fission, the strongest effect was achieved for me in the Ronald Bladen, Nancy Grossman, Cady Noland gallery. Everything here was screaming power, sex and violence. The entire space felt hard as nails—most all of it a macho silver and black. Bracketing the huge gallery were long rows of Nancy Grossman's famous black-leathered heads, aggressively sprouting phallic shapes like picks and horns. Bladen's 1969 minimal masterwork, *The Cathedral Evening*, aggressively dominated the interior space with a mammoth triangle breach. This was backed up by his famous work, *Three Elements* (1965). Then, giving the gallery a sense of an almost palpably Oedipal contest [284] was a large group of superb black-on-silver Cady Noland anthropological silk-screens on metal.[285]

The other room that really collectively worked for me held Paul Thek and the mysterious yet suave Emma Kunz (1892-1963). Three wonderful Paul Thek *Meat Piece* are there; marvellous, weird post-minimal sculptures that sickly encase flayed body sections in wax in long yellow transparent plexiglas shrines that literally shine—suggesting an odd passion for eccentric alternation between lassitude and enthusiasm. This meat-machine mix is counter-pointed with the healing magnetic-field ephemerality of Emma Kunz's geometric drawings, done with lead and colored pencils (or chalk) on graph paper. It was easy to envision some fierce spiritual forces zapping each other without inhibition throughout that room.

Other rooms brought the link-up to a jolting halt. I simply admired Martin Boyce's huge neon sculpture (Boyce channeling the maîtres, Dan Flavin), but it produced no associative noise effects with what else was in the room. Worst of all was a room entirely devoted to the work of Joe Brainard. What was that doing there? One strains to see (or imagine) even a second mind in that space. So the unavoidable thought arises, well, Rondinone must like this stuff—so that is at least two minds in synch. But does Rondinone think there is anything still interesting or perturbing in a Gober sink? His *The Split-up Conflicted Sink* from 1985 also played a huge flat note for me in this supposed visual noise symphony, as did the overly unembellished black crosses of Valentin Carron, the stupid car bashed installation by Sarah Lucas, and the cloying faux-naïve canvases of Karen Kilimnik. How to connect this boring, stupid and naïve work to the third mind connectivity theme then?

Nevertheless, I will. On thinking about the show on my way home, I concluded that the show's thwarted relationship to connectivity is gravely naïve and passé (if pleasant in a quaint, charming way) in lieu of the multi-networked world in which we now reside. By now, various theories of complexity have established an undeniable influence on cultural theory by emphasizing open systems and collaborative adaptability. One ponders if Rondinone has ever even heard of the theories of Tiziana Terranova,[286] Eugene Thacker or other cultural workers involved in the issues of human-machine symbiosis as interface within our inter-networked media ecology. So yes, part of the pleasure for me was bathing in this old-fashioned naïvety, having just spent some serious time reading and writing on the topics of conspiratorial shadow activities and viral software logic based on complex inter-connectionism. Placed against issues of avant-garde cybernetics, the coupling of nature and biology via code, media ecologies, distributed management teams, Internet mash-up music, artificial life swarms, the political herd mind, and Negri/Hardt's multitudes, *The Third Mind* played in my mind like a romp through a kindergarten playpen. Nice infant noise. It felt good to forget about that pervasive nagging political/cultural feeling of stalemate created by the resilience of our current reality in that it assimilates everything.

But no, Ugo Rondinone did not randomly cut and reassemble art to create a new third meaning. He did not cut up anything. Just like every dj, fashion designer, and group show curator, he remixed contemporary expression from recent decades to permit new meanings to emerge. The ideas in the collaborative work of Brion Gysin and William Burroughs were not needed to achieve this end—and perhaps they were poorly intellectually served here (even though it was great to see the work). There was no use of chance or randomness evident here (even the re-shuffled catalogue pages I heard were rather suspiciously non-random) that is necessary for a really unexpected—and perhaps disastrous—result. This show did not go that far. There was no random reassembling of various fragments of something to give them a completely new and unexpected meaning (like I saw in the show *Rolywholyover: A Composition for Museum by John Cage* at the Guggenheim Museum in Soho, NYC in 1994). *The Third Mind* is just a standard, but good, heterogeneous art show where the whole is greater than its parts. Which is as it must be.

Anyway, to get back to my main point: as the blending between the artist and spectator took on greater and greater emphasis in the late-1960s, new forms of aesthetic immersion into noise opened up. It is precisely in this third mind blending that the question of *art as noise ambience* arises. Indeed, *ambience as art is a fruitful domain in which to find the art of noise aesthetic in all of its varieties and forms of manifestation.*

The term *ambience* used here follows Frank Popper's definition of the *artistic environment* as a meeting ground of physical and psychological factors, which implicate the spectator's inherent participation in the art's fulfilment in a delicate, atmospheric way.[287] This is indeed the case with La Monte Young's and Marian Zazeela's *Dream House*: a fully immersive light and subtle sound environment in which the visitor may move about and thereby participate in the formation of the noise-sounds and optical effects encountered. An *artistic ambient environment* is a key concept for aesthetic immersion into noise and we shall return to it.

In *Art, Action and Participation,* Frank Popper showed (with particular reference to post-kinetic research) the convergence and specificity of the notions of environment and creative participation which combined to form the principal direction of art research in the theoretical and practical domains. In *Art, Action and Participation,* a source book from which I drew many examples from this period, Popper found that mixed-media expressions that involve all the senses are conducive to the more complete involvement of the spectator, and that science and technology can act as creative stimulants. In terms of artists of the 1960s working in this new expanded-field, a good example is GRAV (Groupe de Recherche d'Art Visuels), created by Jean-Pierre Vasarely aka Yvaral, son of Victor Vasarely—a group of eleven artists who picked up on Victor Vasarely's concept that the sole artist was outdated. GRAV was active in Paris from 1960 to 1968. Their main aim was to merge the individual identities of the members into a collective and individually anonymous activity linked to the scientific and technological disciplines based around collective events called *Labyrinths*.

Their ideals incited them to investigate a wide spectrum of kinetic and optical effects by using various types of artificial light and mechanical movement. In their first *Labyrinth,* held in 1963 at the Paris Biennale, they presented three years' work based on optical and kinetic devices. Thereafter they discovered that their effort to engage the human eye had shifted their concerns towards those of spectator participation, a foreshadow of interactivity. On April 19th, 1966, GRAV created *Une Journée Dans la Rue* (Day in the Street) in Paris where they invited passing participants to involve themselves in various kinetic activities such as having them walk on uneven blocks of wood and/or experience a distorted world by wearing elaborate distorting spectacles. Their agreed dissolution in November 1968 was based on their recognition that it was impossible to maintain the rigor of a joint program.

Relevant here is the *Nicolas Schöffer Exhibition* I saw in Paris in fall 2005, for if one discounts the existence of László Moholy-Nagy's (1895-1946) Bauhaus *Light Space Modulator* (1923-30)[288] —a visionary multimedia noise artwork that helped inaugurate the artistic dialogue between machines, light, shadow and motion—there is something to the claim that the Hungarian-born French artist Nicolas Schöffer (1912-1992) is "the Father of Cybernetic Art". At the very least, this premise was entertained while viewing actual work (mostly mobile sculpture under theatrical lighting effects) and an incredible amount of documentation at the museum of the French electricity company Espace EDF Electra.

What is immediately evident in this exceptional historic presentation is that Schöffer's career touched on painting, kinetic sculpture, architecture, urbanism, film, TV, and even music (he collaborated with Pierre Henry)—all in the pursuit of a noise dynamism in art which was originally initiated by the Cubo-Futurists and then intensified and solidified by the Russian Constructivists such as Naum Gabo (1890-1977), Anton Pevsner (1866-1962) and Moholy-Nagy. All were concerned with opening up the static three-dimensional sculptural form to a fourth dimension of time and motion, and this was Schöffer's intention as well. Schöffer however, coming well after, benefited pleasingly from cybernetic theories (theories of feedback systems (interactivity) primarily based on the ideas of Norbert Wiener) in that they suggested to him artistic processes in terms of the organization of the system manifesting it (*e.g.*, the circular causality of feedback-loops). For Schöffer, this enabled cybernetics to elucidate complex artistic relationships from within the work itself.

His *CYSP 1*, from 1956, is considered the first cybernetic sculpture in art history in that it made use of electronic computations as developed by the Philips Company. The sculpture is set on a base mounted on four rollers, which contains the mechanism and the electronic brain. Small motors located under their axis operated the plates. Photoelectric cells and a microphone built into the sculpture catch all the variations in the fields of color light intensity (and sound intensity). All these changes occasion reactions on the part of the sculpture.

Consequently, Schöffer's kinetic sculptural compositions were able to parallel Warren McCulloch's (1898-1969) adaptation of cybernetics in formulating a creative epistemology concerned with the self-communication within an observer's psyche and between the psyche and the surrounding environment. This is cybernetics' primary usefulness in studying the supposed subject/object polarity in terms of artistic experience. That is the theoretical premise, at least.

In actuality, I was treated here to dramatic light shows (some on the trippy side) that come whirling out of his spinning mechanical metal sculptures. Colored lights bounce off revolving polished metal towers, casting ever-changing lights and shadows onto huge wall screens and into my eyes. There also was a very basic interactive room consisting of a group of smaller whirling sculptures which responded to my presence and a large prismatic triangle structure containing infinity views.

In Schöffer's triangular structure, my image was ceaselessly mixed and reflected within spinning lights. As such, I was made to feel an integral part of an exploding noise. In general, this infinity noise experience invited me to view myself in infinity, and so to feel space not in the traditional passive Euclidean custom, but in a conceptually operative and viractual (*viractive*) manner.

In addition, the exhibition demonstrated Schöffer's three period styles. First, his "spatio-dynamic" constructions from 1948 on: attempts at a synthesis of spatial and dynamic elements. Next came the "lumo-dynamic" constructions of 1957, which connect light projections to music. In his "chrono-dynamic" works of 1959, word and tone, movement and space, light and color all form together a sum of space-time noise. Also well documented was Schöffer's 52 meter high *Cybernetic Tower* from 1961, which was constructed in Liege with 66 revolving mirrors.

Given the period-piece nature of the exhibition, I found it stylistically engaging in terms of noise art and not overly retro-looking. Indeed, the show surprisingly did not appear all that dated, even though of course it recalled the early Paris 1960's and the futuristic "space age" designs of Paco Rabanne, which involved the use of moving metallic discs or plates. Yet my subject/object polarity never shifted much.

But given this, shouldn't Nicolas Schöffer's work be considered something other than an art object per se? Perhaps it is more appropriate to think of it as a means of transforming static perspective vision into a luminous study. We might just as well consider it then as stage props. Or better, an apparatus for painting with light.

With his video works of 1961, Schöffer is additionally regarded as an early representative of video art—so perhaps it all funnels into special effects broadcast TV (which he did). For me, the final interest of this show (which I saw three times) is in its allowing me to better position Schöffer in a certain art-tech artist-engineer intellectual history—a living history that has not yet exhausted itself. Indeed it is touching to consider that László Moholy-Nagy's *Light Space Modulator*—which was driven by a motor and equipped with 128 electric bulbs in different colors—was finally demonstrated at the 1930 Paris Werkbund exhibition. So I see Nicolas Schöffer here not only as a pioneer of cybernetic art, kinetic sculptor, town planner, architect and theoretician of art, but as a key player in the middle of the art-tech intellectual narration—a narration that increasingly defines artistic achievement in the beginning of the 21st century.

Also significant in immersive noise terms from that period is Stan Vanderbeek's 1966 *Movie Drome*, a hemispherical "movie-mural" created in upstate New York State where the viewer assumed a supine position to look upon an onslaught of hemispheric cinematic projections.[289] As Vanderbeek himself described it, the Movie Drome operated as follows: "In a spherical dome, simultaneous images of all sorts would be projected on the entire dome-screen. The audience lies down at the outer edge of the dome with their feet towards the center, thus almost their complete field-of-view is the dome-screen. Thousands of images would be projected on the screen".[290] According to Vanderbeek, details of this hour-long "multi-plex" dense image flow (inherently excessive) were not important. What was important was a "total scale" felt in rapport with the "rapid panoply" (what Vanderbeek called the dome's "visual-velocity") which functioned so as to "penetrate to unconscious levels".[291] This hemispheric reconfiguration of the screen (so as to heighten film's immersive appeal in terms of filling the FOV) conforms to what Jonas Mekas called *absolute cinema*.[292]

In addition, Francis Thompson, best artistically known for a six screen projection arrangement called *We Are Young* which covered a total area of 885.6 square meters (2,952 square feet) at the Expo '67 in Montreal, produced large-scale immersive projections based on his interest in having films optically swallowing an audience. Thompson said about these large displays that he "would like to see a theater with so great an area that you no longer think in terms of a screen: it's the *area* you're projecting on". Then images would "come out of this surrounding area and hit you in the eye or go off into infinity. So you're no longer working with a flat surface but rather an infinite volume".²⁹³

Non-immersive noise cinema makes use of what is called *framing*. Framing is intended to eliminate what is deemed unessential in the motion picture, to direct the spectator's attention to what is important and to give it special meaning and force. Each frame of film, which corresponds in shape to the image projected on the screen, forms the basis for a graphic composition in the same way as the frame of a painting encloses the area in which the painting must be organized. Several different ratios of frame width to frame height (called *aspect ratios*) have been used in motion pictures. The most common, known as the Academy ratio, is 1.33 to 1, or 4 to 3, a ratio corresponding to the dimensions of the frame of 35 millimeter film. By using 70 millimeter film or a special CinemaScope lens, an image with wider horizontal and shorter vertical dimensions is achieved; a proportion of about 5 to 2, or between 2.2 to 1 and 2.65 to 1. A similar effect, called wide-screen, was sometimes achieved without the expensive equipment required for CinemaScope by using 35 millimeter film and masking the top or bottom, or both, giving a ratio of 1.75 to 1, or 7 to 4. Although some theaters in the 1970s were enlarged and widened to accommodate 70 millimeter images, a trend toward smaller theaters fixed the image ratio close to 1.85 to 1 in the United States and 1.66 to 1 in Europe.

Rejecting the framing trope for art, in 1954 Yaacov Agam began to undertake research into what he called *transformable structures* (the equivalent of paintings and reliefs) and *transformable objects* (the equivalent of sculpture) where the spectator was obliged to take up successive positions in front of the reliefs in order to discover the sequence of changing lines, forms, colors and structures which offered themselves from different exclusive angles. Agam himself pointed out that all his works are in fact *transformable*, but he reserves the term in particular for those in which the basis of the transformation lies in being able to modify the pictorial structure, for example in the 1953 piece, *Nuit*. He extended this premise immersively with his *Total Picture Environment Salon* at l'Eysée in Paris.

According to Gene Youngblood, with the art of Marcel Duchamp, John Cage, and Andy Warhol (1928-1987), western civilization "rediscovered art in the ancient Platonic sense in which there's no difference between the aesthetic and the mundane".[294] This, we can say, is the basis of Pop Art (a term coined in 1958 by the critic Lawrence Alloway) as Pop Art found its imagery and many of its techniques in the realm of advertising and consumer packaging and pop stars and cinema idols.

Most definitely, the Pop-Happenings of Andy Warhol's art-music group, the Exploding Plastic Inevitables (E.P.I.) (which eventually became the rock group The Velvet Underground) is the most conspicuous Pop noise work, as the audience and the players/performers were embedded in a high volume light/sound/film show which dominated the space and stirred the consciousness of those watching or dancing. E.P.I. Happenings first were performed in the spring of 1966 at a Polish dance hall on St. Mark's Place in New York City called *Polsky Dom Narodny*. Warhol rented the *Dom* (home) from two artists who "sculpted with light", Rudy Stern and Jackie Cassen, and painted it white so that movies and slide projections could be cast on the walls in wallpaper-like fashion. Five movie projectors were utilized along with five carousel-type slide projectors which could each change an image every ten seconds. The slides were projected directly onto the films, whose sound tracks would sometimes be played, and thus blend in with the live music/hullabaloo. A mirror-ball also was utilized along with spot-lights and strobe-lights.[295]

E.P.I.'s noise Happenings aimed to achieve a traumatically dazzling ontological restructuring of consciousness. Here the space of the Happening (light-show/concert/film-show/live-performance) verges on the all-consuming in a way now familiar to those who have participated in techno-raves, rock concerts, and/or house music clubs (such as the legendary Paradise Garage (1976-1987) in New York City, a club that attained an added immersive noise sweep to its milieu by embedding powerful sound-speakers under its dance floor). Indeed, the now ubiquitous mirror-ball (whose inventor I was not able to uncover) must be recognized as an immersive noise artwork of significant stature.

In 1969, a 210° immersive noise Gesamtkunstwerk model was first created by the Los Angeles wing of the Experiments in Art and Technology (E.A.T.)[296] project for the Pepsi-Cola Pavilion at *Expo '70* in Osaka, Japan. The 210° mirrored sphere, whose prototype was shown in the U.S.A. in a Santa Ana blimp hanger in September 1969, was simply a light-weight mirrored sphere constructed from 3,900 square meters (13,000 square feet) of mirrored mylar 2,540th of a centimeter thick

(1,000th of an inch) which spanned 27 meters (90 feet) in diameter and 16.5 meters (55 feet) in height.[297]

Another Pop-Happening artist emblematic of an all-consuming immersive noise aesthetic is the previously mentioned Japanese artist Yayoi Kusama, who in 1958 moved for a time to New York City. Kusama's dominant obsessions have been the excessive accumulation of polka-dots (or extraordinary milky phallic growths) which on occasion span entire rooms and create very noisy experiences. As a juvenile in Japan, Kusama developed an obsessive-compulsive disorder that caused encircling hallucinations, as mentioned previously. Kusama's obliterating installations strive to characterize a waking hallucinated-vision she experienced as a young woman, where sitting at a table covered with a floral tablecloth, in a room covered with floral wallpaper, she saw that her hands, too, were covered with flowers. As she herself has said, "There was a vase of golden violets and when I looked at them and then looked away they began to cover everything. They were on the drawings I was doing and then I saw that they were all over the phone book, and going up the walls and then they covered the doors so that I could not see a way out of the room. These experiences were typical".[298] As a result, it is as if Kusama was agonizing to overwhelm the entire world with the noise of her polka-dots, and it is this explosive and immersive-noise model that places her at the forefront of the noise-Pop post-war avant-garde, reflective, as her installations, happenings, and even paintings are, of noisy psychological obsessions.

Kusama's noisy *Infinity Net* paintings are comprised of small, overlapping, looping brush strokes which create optical-fields reminiscent of a boundless sea. This, of course, ties her paintings into issues of the previously discussed painterly noise sublime. Moreover, she also has created an extraordinary group of *Net Paintings* in the late-1950s and early-1960s (for example *Yellow Net*) which insinuate the overall webbed-infinite hyper-total aesthetic accomplished by Jackson Pollock. This implied immersive-overall noise hyper-space becomes literalized in her infinity cubed mirror installations at the end of the 1960s and early 1970s.

As she has said, "In New York, I was painting the red nets and then I noticed that it spread to the floor and the curtain and to the window. So I went to catch the red net, and I examined it without noticing at first that my hands were also covered by the red nets. And that was the turning point, and I started creating sculpture, so that I could put the patterns on everything".[299] This desire is realized in the installations *Repetitive Vision* and *Dots Obsession*. In *Dots Obsession*, a room 4.8 meters wide by 15 meters long by 3 meters high (16 by 50 by 10 feet) had been painted an intense yellow with different sized black dots randomly placed on the walls, floor, and ceiling. Three huge, organically-shaped balloons (one was 9 meters long by 3 meters high (30 feet by 10 feet)) are the same color as the room, right down to the black dots that filled the total space.

Her grottoesque noise installation *Repetitive Vision* was approached by walking first into a black corridor and then into an intensely lit space whose floor was covered in hot-red dots. I encountered there three female mannequins painted white, their bodies and hair covered with the dots, reflecting to infinity in the mirrored walls and ceilings.

As Kusama is consistently motivated by her desire for an obliteration of the self in visual-noise-infinity characterized by the all-over use of polka-dots (so immersive is this impulse that Kusama often covered her skin and hair in polka-dots), her mirrored immersive installations are salient noise sites in which to explore issues of disembodiment (issues of self devastation of cognitive self-body-image) and willed visual self-obliteration, as when within them the viewer may merge with, and dissolve into, the visual panorama reflected *ad infinitum* in the walls of mirrors. The effect is as if being itself was being circuitously inhaled.

To immerse one more fully in her proliferating noise environments, in 1965 Kusama turned to the use of mirrored-rooms to enhance the feeling of expansive immersion into noise *ad infinitum* with the construction of *Narcissus Garden*, *Kusama's Peep Show* and *Endless Love Room*, for example.

Stylistically, this work can be seen as a synthesis of Op, Pop and Psychedelic Art, and there is the obvious communality she shares with Lucas Samaras' 1966 *Room 2* and Christian Megert's environments (which also incorporate mirrors) as in the *Spiegelraum* that was included in the *Environments* exhibition in Utrecht in 1968 and in *Mirror Environment* included in Documenta 4, Kassel. Moreover, though less immediately all-encompassing, but perhaps even more highly charged with total noise symbolism applicable to the entire environment, in 1969 Robert Smithson (1938-1973) began producing works in the landscape called *Nine Mirror Displacements*, *Mirror Shore* by placing mirrors on a beach or in the jungle of the Yucatán. Smithson then took photographs of these ordinary mirrors set out on the ground and what they were reflecting back.

The Italian artist Getulio Alviani, in his 1964-9 *Cubic Environment,* also made use of reflective media towards immersive noise ends on an architectural scale, and again in *Surface and Texture* (1969), an aluminium wall 3.20 by 5.60 meters in size (about 10.5 by 18.4 feet). His *Interelazione Cromospeculare* was wonderful to experience, as I did, at the *Italics: L'art italien entre tradition et révolution 1968-2008* show in Venice at the Palazzo Grassi in 2008. Luc Peire's *Environment* was also constructed as an enveloping reflective arrangement where mirrored surfaces rebound amplitude to an indefinite degree in order to help one achieve noise consciousness of the unlimited dimensions of vibrant space.

In like manner, Domingo Alvarez created mirrored rooms out of a number of large mirrors in an entirely closed construction which projected a self-conscious space outward *ad infinitum,* as in his 1972 *Mirror Environment.* Its seemingly immeasurable space might seem exclusively and merely external to one at first glance, but if considered closely, penetrates consciousness noisily. We saw this in Kusama's installation *Infinity Dots Mirrored Room* where a white Formica floor was covered with three sizes of colored fluorescent dots within a mirrored-room teeming with blacklight. By being ceaselessly reflected on the ceiling and walls, one felt an integral part of the exploding noise. In general, then, the infinity noise experience bound me to view and feel space not in the traditional passive custom but in a conceptually operative and viractual (*viractive*) manner.

It is partly for this reason of noise visuality, I submit, that the use of mirrors in art flourished during the late-1960s, curiously around the same time the post-structuralist French psychologist Jacques Lacan (1901-1981) was broadly publishing his theory of the *mirror stage* in human development in the chapter titled "The Mirror-Phase as Formative of the Function of the I" in *Ecrits* (originally published in 1949 in French). The Lacanian term *mirror stage* indicated the point in a child's growth when the psychological feeling of undifferentiated unity with the mother is substituted with a conception of a disconnected self. According to Lacan, the experience of perceiving oneself in a mirror, literally or figuratively, generates internal trepidation (noise) inasmuch as one *anticipates and wills for oneself a homogeneous total being* over which the ego has dominion. However, this totality is never achieved, so that one's spellbound ego comes to feel inadequate.[300]

Moreover, Lacan emphasized the primacy of language as the mirror of the unconscious mind, and he tried to introduce the study of language (as practised in modern linguistics, philosophy, and poetics) into psychoanalytic theory. His major achievement was his reinterpretation of Freud's work in terms of the structural linguistics developed by French writers in the second half of the 20th century. The influence he gained extended well beyond the field of psychoanalysis to make him one of the dominant figures in French cultural life during the 1970s, and in Critical Studies within Anglo-Saxon academic circles from the early 1980s on.[301]

Coming at this noise mirror issue from an almost polar-opposite position is the American artist Bruce Nauman's 1968 efficacious noise installation called *Get Out of My Mind, Get Out of This Room*, that consisted of an empty, small, white room, filled only with noise that seems to come from all directions. Simply constructed, it consisted of loud-speakers invisibly embedded into the walls which played a male voice shouting and moaning the injunction of the title. There is nothing to see, yet the rhythmic pattern of the voice bleating out this repetitious ornately coupled incantation without end locks one into a surround-sound cognitive/dissonant noise situation of attraction/repulsion. However, if we are not to settle for affirmations of the emptiness of our being, it seems to me that any noise art proposition must also be an initiatory one done at the limits of ourselves which must, on the one hand, open up a realm of ontological doubt, but on the other, put itself to the test of affinity with contemporary ideas of infinity, both to grasp the points where noise expansion is possible and desirable, and to ascertain the accurate form the expansive proposition should take. This means that the constructed ontology of ourselves must turn away from all projects that claim to be determined and restricting by persisting in an immersive noise consciousness both rhizomatic and infinite.

In terms of the immersive noise art of the 1960s which addressed contemporary concepts of the infinite, mention should be made of the Dvizjenije movement in Moscow and its leader Lev Nusberg. The Dvizjenije movement adapted the "cosmic ideas of the Malevich tradition" [302] in an attempt to construct what were called *Living Machines* (i.e., kinetic environments) between the years 1962 and 1967. Lev Nusberg himself had in effect been working since 1964 through 1967 on noise projects concerned with setting up *artificial kinetic milieus* which would register kinetic sensations within them, though only partly realized. One Nusberg project idea was for a kinetic labyrinth that was to extend 500 meters (1,640 feet) and branch off into several different directions containing a large number of consecutive shaped and colored rooms accommodating, at various points, film, music, mime performance, text, kinetic objects, smells, and even air currents. This atmospheric approach to art is also evident in Carlos Cruz-Diez's environmental color-events called *Chromosaturations* where atmospheric three-dimensional color experiences were encouraged in various rooms or booths, thus bringing one in direct contact with a unique sensory encounter by hanging homogeneous color in space. In some of his *Chromosaturations*, the visitor, after being "decontaminated" in transitional coal-black chambers, passed through a sequence of consecutive chromatic situations in which one experiences sheer blue, red, and green. In his *Chromosaturations for a Public Place in the Open Air,* exhibited at Venice, Cruz-Diez returned to an idea which he had already put into effect at the Carrefour de l'Odéon (Paris) in 1969 where pedestrians were invited to enter and pass through a series of differently colored-filled booths. In the version exhibited at Venice, this principle was carried further by inducing the spectator to follow a corridor of continuous color saturation so that one successively experienced absolute blue, absolute red and absolute green. Cruz-Diez thus achieved a *total vision*[303] through the summation of distinct monochrome perceptions. Mathilde Perez also created complete color experiences by constructing a prolonged corridor of unified chroma which essentially brought one into the experience of pure color carefully modified in such a way as to permit the sensory perception of colored space.

This noise emphasis on art as "a kind of sensory-stimulation laboratory"[304] took on the cybernetically charged open-field and provided inputs for a post-modern noise activity generally characterized by a *process aesthetic* and a *de-objectification* that emphasized the artist's encounter with the palpable and malleable properties of reality from within the conglomerate atmosphere. In the open-field, "sculpture" came to incorporate wholly new modes of compositional events, such as earthworks and media art: film, video and electronics. Various conditions of presentation (including site-specific installations and street works) brought art further from the framework constraints of the picture frame and the traditional function of the gallery and deeper into noise. In Process Art, Conceptual Art, and Earthworks, there was a sense of common motivation: an effort to escape the conventional terms of the art object as nurtured by the museum/gallery milieu and to move art out into a broader context. Here was a definite opening towards the noisy environment, coupled with an appeal to general creativity that was evident from the very simplicity of the materials and statements.

Another aspect of noise practice in the expanded-field is that of the artist becoming his or her own work of art, totally losing the usual boundaries between 'art' and 'life' and 'artist' and 'work'. This tendency is best represented by Linda Montano, the founder of the Art/Life Institute and the main defender of Living Art. Living Art is an attempt to merge art and lifestyle through long-term performance works, defined as any work/play that artists/non-artists are willing to perform together or alone. Montano's Living Art performances, which she has created over the past 25 years, include *Three-Day Blindfold* (1975) and a co-operative work with Tehching Hsieh in which the two artists spent a year tied together by a 2.4 meter (8 foot) rope (1983-84). This work had the additional stipulation that the artists not touch.

The 1981 immersive noise performance collaboration between Bill Seaman and Carlos Hernandez called *Architectural Hearing Aids* touches on this immersive Living Art mode in a noisy way, as it drove the participant in a car installed with two different sound systems and a 4-track mixer and seven speakers on a specific tour of San Francisco. Sound/music was composed specifically to alter perceptions of the real architectural structure of the city.

Noisily, Joseph E. Furey's (1906-1990) Brooklyn railroad apartment at 447 Sixteenth Street was completely covered with brightly painted cardboard appliqués, shells, and other found objects so that the walls were teeming with stippled dots of black, green, beige and red paint that covered thousands of clam shells and hand-cut cardboard hearts, cross shapes, and diamonds. Mussel shells, spread open to resemble butterflies, were bordered by colored tile and chips of mirror, lima beans, and glass beads. Bits of collage, pictures of monkeys, butterflies, and dogs, dotted the wallpaper landscape mural.

There, too, is ST EOM's *Pasaquan*, created between 1958 and 1984 by Eddie Owens Martin (1908-1986)[305] located in the sand hills of south-west Georgia, near the small town of Buena Vista. Pasaquan consists of eight grass covered acres, surrounded by pine trees and adorned with walls, pagodas, buildings, temples, and walkways—all brilliantly decorated, inside and out. This all-over noise tendency to address a space aesthetically has merited the term Installation Art, an artform I will illustrate with Milton Cohn's late-1960's *Space Theater*. The essence of Cohn's *Space Theater* was a rotating assembly of mirrors and prisms adjustably mounted on a flywheel around which were arranged a battery of light, film, and slide projectors. Essentially, *Space Theater* was an expanded version of Moholy-Nagy's *Space-Light Modulator* into which one may enter. Cohn's intention was to "free film from its flat and frontal orientation and to present it within an ambience of total space".[306]

This desire to create a surrounding projective noise space was also the intent of Jud Yalkut's late-1960s *Floating Theater*. The *Floating Theater* consisted of a parachute canopy 9.6 to 15 meters (32 to 50 feet) in diameter anchored with nylon strings on which projections were cast. The canopy, which reflected both rear and frontal multiple-projections, was suspended over and around the audience by the use of fans.[307] Henry Jacobs and Jordan Belson's intermedia *Vortex Concerts*, realized intermittently between 1957 and 1960 in the 18 meter (60 foot) domed Morrison Planetarium in San Francisco's Golden Gate Park firmly established the surrounded/expanded field-of-view objective as well. As reported by Youngblood,[308] the *Vortex Concerts* consisted of an abstract light/sound presentation created with "interference-pattern projectors" that intermixed with the surrounding star projections and strobe lights, an experience which engulfed the onlooker in art noise. Belson explained that in the *Vortex* they were able to "project images over the entire dome so that things would come pouring down from the center, sliding along the walls".[309] Also, Aldo Tambellini's mid-1960s environmental *Electomedia Theater* served as a robust precursor to this surrounded/expanded noise trend. Particularly significant are Tambellini's 1965 *Black Zero* environment, which engulfed the viewer in a "maelstrom of audio-visual events",[310] and his 1968 collaboration with Otto Piene on the multi-channel, closed-circuit environment called *Black Gate Cologne*, a performance environment that helped re-establish this surrounded/expanded noise inclination again in Europe.

This appetite to fill us in projection noise was appeased in inverse micro-fashion in a one-on-one interactive assisted film-performance by Bradley Eros called *Movie Head Box* which he presented as part of *The Extremist Show* at ABC No Rio in New York City in 1983. Eros provided me with a screen-box which slid over my head (like a primitive HMD). He then projected a color super-8 film onto my head-screen's façade (which insinuated a noisy erotic chronicle dipped in alchemy) thereby over-flooding my visional capacities, as the erotic/alchemical images seeped into the box and were reflected off its inner sides. A walkman provided an intimate noise soundscape[311] accompaniment made up of metallic abstract sounds.

All of the above examples demonstrate that there has been a developing taste for what I have identified as *immersive noise art*: art that attempts to project its aesthetic noise value ambiently but coherently throughout an expanded aesthetic field-of-view.

CHAPTER 5

Viral Attack within Connectivist Noise Schematics

{loop:file = get-random-executable-file;
if first-line-of-file = 1234567 then goto loop;
prepend virus to file;}
— Fred Cohen, *Computer Viruses: Theory and Experiments*

We cannot be done with viruses as long as the
ontology of network culture is viral-like.
— Jussi Parikka, *The Universal Viral Machine*

Dada is the viral option to the virtual certainty.
— Andrei Codrescu, *The Posthuman Dada Guide*

We have seen that the aesthetic logic of noise may no longer be reduced to the unwelcome. What was said of the subtle and peculiar noise of the Rococo interior—and its suggestive resemblance to the vast array of nerve bundles descending from the cortical areas onto the intralaminar nuclei and the nuclear reticularis in the thalamus and its array of massively inter-connected neural circuits—is also expansively applicable to the all-over interlacing network of today: the World Wide Web.

Let's review how we got here. On October 27th, 1969, two computers began exchanging messages with each other through a link leased from the telephone company as part of an experiment funded by the United States Department of Defense Advanced Research Projects Agency (ARPA). Researchers did so seeking to construct a resilient inter-network of military communications that could survive the destruction or failure of any, even most, of its component parts and maintain communication in a nuclear war. Hence the Internet was designed to have no center and limited hierarchy by, ironically, the most hierarchical of institutions known to democracy, the US military/industrial complex. In reviewing these simple facts, I am assured that my previous proposition that the noise of militarized space and the threat of annihilation of the civilian population has been the primary motivational force responsible for the boom in immersive art noise following the Second World War. Indeed, I can say that this narrative of militarized space will continue to be the unconsciously encoded noise impulse in the future. Everything, everywhere, all at once in a rhizomatic web of communication: this is (and will continue to be) the zeitgeist (spirit of our age) inherited from the C3I military model which, according to Hegel, ensures a similar complexion in all the activities of a period, from science to art, literature and music. Hence intricate lattices of linked noise, available to all at any time, characterizes my hypothetical understanding of where noise culture is today.[312]

But noisily, the web also regenerates deep connections to the past; so cyberspace, this territory which stretches out from hypertext to the world-wide computer network, from virtual reality to video games, might also be theorized as the domain of Raymond Roussel's (1877-1933) idea of noise without place, reduplicating without duplication, reiterating without repeating. As with the conceptual machine works of Roussel, viral web art is a coldly concerted and particularly dizzying activity.[313] It is a strident activity lost in an infinite navigation from one sort of encounter to another in which the affirmation of the other keeps appearing and disappearing in the play of mechanical maneuvers (or mechanisms) destined to avert gratification. This is where the bachelor apparatus of Duchamp repeats itself ad infinitum by transmitting the noise of the machine via alter-ego. For example, such an art of noise is in almost all of Knowbotic Research KR+cF projects that I have encountered,[314] but mostly in their work *Simulation Mosaik Data Klaenge* from 1993. Knowbotic Research KR+cF (principally Yvonne Wilhelm, Christian Hübler and Alexander Tuchacek) have experimented with so called *intelligent agents*—applications that can conglomerate diaphanous information by themselves (also called *knowbots*) and intelligent virtual spaces (flexible information-environments distributed in electronic networks). Theirs is noise art spread wide.[315]

It seems to me that network art noise[316] like theirs will continue to develop into a transforming endeavor affecting the full spectrum of culture. Such networked art noise will of course be multi-national, which in itself implies a growing super-totality unfettered by many physical limitations (once the required technological hard- and software is at hand). The enlargement of this linked art noise is only (fundamentally) limited by lack of human imagination, lack of equipment and the knowledge to use it, and by what is numerically or mathematically feasible. Hence I think it fair to say that the possibilities that noise art offers our net ecology today are enormous.

Noise art, I hope I have demonstrated by now, is about being in the ready position of excess, and the Internet's World Wide Web, of course, is the means for linking the excess of noise art. On the Web, information can be smoothly accessed in a synchronous system permitting anyone connected to click and enter. This affects the speed with which new associations are assembled and disrupted, as well as the kinds of interactions that arise and emerge, which together allow art noise to be linked in terms of desire rather than physical geographic position. This net-condition allows new feedback-loops of noise theory and noise experimentation not formerly obtainable to emerge. New noise artists and their Internet noise music can, through this net-condition, become more accessible, permitting a closer aesthetic symbiosis between computer technology and culture. Assuredly, with the conflation of noise art and the World Wide Web (which strings such art together), noise artists better procure the connectivist-perspective of the network that Roy Ascott has identified and encouraged.[317] Hence connected noise art can advance a net-condition awareness of rupture and plurality in hyper-homogeneity (a supplementary order of diversity within orders of hyper-noise) as noise puts us in the position of initial *critical distance.*

Again the militarization (and subsequent de-militarization through art) of consciousness is what will be fashioning this net-conditioned scenario into an eventuality when linked to the forces of capitalist colonization.

Our technology is historically informed not only by its materiality but also by its political, economic, and social context. Even so, the uncommon visionary artist may override these tendencies by envisioning discernibly different utilizations of the technology via noise.[318]

To investigate how the art of noise applies to linkage today, I will now discuss the book, *Digital Contagions: A Media Archaeology of Computer Viruses,* by Jussi Parikka. One could be forgiven for assuming that a book with that title would be of sole interest to those sniggering horn-rimmed programmers who harbor an erudite loathing of Bill Gates and an affection for the Viennese witch-doctor. Actually, it is a rather game and enthralling look, via a media-ecological approach, into the acutely frightening, yet hysterically glittering, networked noise world in which we now reside. A world where the distinct individual is pitted against—and thoroughly processed by—post-human semi-autonomous software programs which often ferment anomalous feelings of being eaten alive by some great indifferent artificiality that apparently functions semi-independently as a natural being.

Though no J. G. Ballard or William S. Burroughs, Jussi Parikka nevertheless sucks us into a fantastic black tour-de-force narrative of virulence and the cultural history of computer viruses followed by innumerable inquisitive innuendoes concerning the ramifications for a creative and aesthetic, if post-human, future. A computer virus is a self-replicating computer program that spreads by inserting copies of itself into other executable code or documents. A computer virus behaves in a way similar to a biological virus, which spreads by inserting itself into living cells. Extending the analogy, the insertion of a virus into the program is termed an *infection*, and the infected file, or executable code that is not part of a file, is called a *host*.

Digital Contagions is impregnated with fear and suspicion, but we almost immediately sense that it also contains an undeniable affirmative nobility of purpose, which is to save the media cultural condition—and the brimful push of technological modernization in general—from catastrophically killing itself off.

This admirable embryonic redemption is achieved by a vaccination-like turning of tables, as Parikka convincingly demonstrates that computer viruses (semi-autonomous machinic/vampiric pieces of code) are not antithetical to contemporary digital culture, but rather *essential* traits of the techno-cultural logic itself. According to Parikka, digital viruses in effect define the media ecology logic that characterizes our networked computerized culture in recent decades.

We may wish to recall here that, for Deleuze and Guattari, media ecologies are machinic operations (the term machinic here refers to the production of consistencies between heterogeneous elements) based in particular technological and humane strings that have attained virtual consistency. Our current inter-network ecology is a comparable combination of top-down host arrangements wedded to bottom-up self-organization where invariable linear configurations and states of entanglement co-evolve in active process. Placing the significant role of the virus in this mix in no uncertain terms, Parikka writes that, "the virus truly seems to be a central cultural trope of the digital world".[319] Indeed Parikka recognizes digital viruses as the crowning culmination of current postmodern cultural trends—as viruses, by definition, are merger machines based on parasitism[320] and acculturation. So it is not only their symbolic/metaphoric power that places them firmly in a wider perspective of cultural infection, it is their formal structure, in that they procure their actuality from the encircling environment to which they are receptively coupled.

Moreover, with the love of an aficionado, Parikka lucidly demonstrates that computer viruses are indeed a variable index of the rudimentary underpinning on which contemporary techno culture rests. He astutely anoints the indexical function of the virus by establishing not only its symbolic melancholy power in relation to the human body and sex, but by folding the viral life/nonlife model into key cultural areas underlying the digital ecology, such as bottom-up self-organization, hidden distributed activity and ethereal meshwork. Scientists have argued about whether viruses are living organisms or just a package of colossal molecules. A virus has to hijack another organism's biological machinery to replicate, which it does by inserting its DNA into a host. In that sense, Parikka describes network ecology as both actual and virtual, what I have previously identified as the *viractual*.[321] But some viruses do not simply yield copies of themselves, they also engage in a process of self-reproducing autopoiesis: they are copying themselves over and over again but they can also mutate and change, and by doing so, Parikka maintains, reveal distinguishing aspects of network culture at large.

I would add that they mimic the manneristic noise aspects of late post-modernism in general, particularly if one sees modernism as the great petri dish aggregate in which we still are afloat. So computer viruses are recognized here as an indexical symptom also of a bigger cultural tendency (noise) that characterizes our post-modern media culture as being inserted within a modern (purist) digital ecology. This aspect provides the book with a discerning, yet heterogeneous, comprehension of the connectionist technologies of contemporaneous techno-culture.

But beyond the techno-cultural relevance, the significance of the viral issues in Parikka's book to all cultural production is evident to anyone who has already recognized that digitization has become the universal technical platform for networked capitalism. As Parikka himself points out, digitization has secured its place as the master formal archive for sounds, images and texts.[322] Digitization is the double, the gangrel, that accompanies each of us in what we do—and which accounts for our cultural feelings vacillating between anxiety and enthusiasm over being invaded by something invisible—and the sneaky suspicion that we have been taken control of from within.

To begin this caliginous expedition, *Digital Contagions* plunges us into a haunting, shifting and dislocating array of source material that thrills. Parikka launches his degenerate seduction by drawing from, and intertwining in a non-linear fashion, the theories of Gilles Deleuze and Félix Guattari (for whom my unending love is verging on obsession), Friedrich Kittler, Eugene Thacker, Tiziana Terranova, N. Katherine Hayles, Lynn Margulis, Manuel DeLanda, Brian Massumi, Bruno Latour, Charlie Gere, Sherry Turkle, Humberto Maturana and Francisco Varela, Deborah Lupton, and Paul Virilio. These thinkers are then linked with ripe examples from prankster net art, stealth biopolitics, immunological incubations, the disassembly significance of noise, ribald sexual allegories, antibody a-life projects, various infected prosthesis, polymorphic encryptions, ticklish security issues, numerous medical plagues, the coupling of nature and biology via code, incisive sabotage attempts, anti-debugging trickery, genome sequencing, parasitic spyware, killer T cell epidemics, rebellious database deletions, trojan horse latency, viral marketing, inflammatory political resistance, biological weaponry, pornographic clones, depraved destructive turpitudes, rotten jokes, human-machine symbiosis as interface, and a history of cracker catastrophes. All are conjoined with excellent taste. The shock effect is one of discovering a poignant nervous virality that has been secretly penetrating us everywhere.[323]

Digital Contagions's genealogical account is proportionately impressive, as it devotes satisfactory space to the discussion of historical precedent, including Turing machines, Fred Cohen's pioneering work with computer viruses, John von Neumann's cellular automata theory (i.e. any system that processes information as part of a self-regulating mechanism), avant-garde cybernetics, human immunodeficiency virus (HIV), the Creeper virus in the Arpanet, the coupling machines of John Conway, the nastily waggish Morris worm, Richard Dawkins's meme (contagious idea) theory, and even the under known artistic hacks of Tommaso Tozzi. Furthermore, the viral spectral as fantasized in science fiction is adequately fleshed out, paying deserved attention to the obscure but much loved (by me, anyway) 1975 book *The Shockwave Rider* by John Brunner and the celebrated cyberpunk novel by Neal Stephenson, *Snow Crash*, among other speculative books and hallucinatory films.

But the pinnacle of interest, for me, of this engaging and educative read is its conclusion where Parikka sketches out an alternative radical media-ecological perspective hinged on the viral characteristics of self-reproduction and a coupling of the outside with the inside typical of artificial life (a-life). He correctly maintains that viral autopoiesis undertakings, like Thomas S. Ray's *Tierra* virtual ecology art project, provides quintessential clues to interpreting the software logic that has produced, and will continue to produce, the ontological basis for much of the economic, political and cultural transactions of our current globalizing world.

Here he has rendered problematic the safe vision of virus as malicious software (virus as infection machine) and replaced it with a far more curious, aesthetic and even benevolent one, as whimsical artificial life (a-life). Using viral a-life's tenants of semi-automation, self-reproduction and host quest, Parikka proposes a living machinic autopoiesis that might provide a Moebius strip-like ontological process for culture.

Though suppositional, he bases his procedure in formal viral attributes—not unlike those of primitive artificial life with its capability to self-reproduce and spread semi-autonomously (as viruses do), while keeping in mind that Maturana/Varela's autopoiesis contends that living systems are an integral component of their surroundings and work towards supporting that ecology. Parikka here picks up that thread by pointing out that recent polymorphic viruses are now able to evolve in response to anti-virus behaviors. Various viruses, known as retroviruses,[324] explicitly target anti-virus programs. Viruses with adaptive behavior, self-reproductive and evolutionary programs can be seen, at least in part, as something alive, even if not artificial life in the strongest sense of the word. Here we might recall John Von Neumann's conviction that the ideal design of a computer should be based on the design of certain human organs—or other live organisms. The artistic compositional benefit of his autopoiesic virality theory, for me, is in allowing thought and vision to rupture habit and bypass object-subject dichotomies.

I wish to point out here that, although biological viruses were originally discovered and characterized on the basis of the diseases they caused, most viruses that infect bacteria, plants and animals (including humans) do not cause disease. In fact, viruses may be helpful to life in that they rapidly transfer genetic information from one bacterium to another, and viruses of plants and animals may convey genetic information among similar species, helping their hosts survive in hostile environments.

Already various theories of complexity have established an influence within philosophy and cultural theory by emphasizing open systems and adaptability, but Parikka here supplies a further step in thinking about ongoing feedback loops between an organism and its environment; what I am tempted to call *viralosophy*. Viralosophy would be the study of viral philosophical and theoretical points of reference concerning malignant transformations useful in understanding the viral paradigm essential to digital culture and media theory that focuses on environmental complexity and inter-connectionism in relationship to the particular artist. Within viralosophy, viral noise comprehension might become the eventual—yet chimerical—reference point for noise culture at large in terms of a modification of parameters, as it promotes parasite-host dynamic interfacings of the technologically inert with the biologically animate, probabilistically.

So the decisive, if dormant, payload that is triggered by reading this book, for me, is an enhanced understanding of pagan and animist sentiment that recognizes non-malicious looping-mutating energy feedback and self-recreational dynamism that informs new aesthetic becomings that may alter artistic output. An example might possibly be a heuristic becoming[325] that transgresses the established boundaries of nature/technology/culture such as the time-bomb cognitive nihilism of Henry Flynt.[326] This affirmative viral payload forces open-ended multiplicities onto art that favor new-sprung conceptualizations and rebooted realizations. Here the artist comes back to life as spurred a-life, and not as a sole articulation of the pirated environment of currency. So the so-called art virus is not to be judged in terms of its occasional monetary payload, but by the metabolistic characteristics that make art reasonable to discuss as a form of extravagant artificial life: triggered emergence, resilience and back door evolution.[327]

How might this approach to noise work in a physical expanded field? Perhaps R&Sie(n)'s 2005 exhibition *I've heard about...* © at the Musee d'Art Moderne de la Ville de Paris's temporary space at the Couvent des Cordeliers suggested to me a possibility. R&Sie(n) is an investigational architectural firm consisting of François Roche, Stéphanie Lavaux, and Jean Navarro; working here with Benoît Durandin. Together, they utilized generative heterogeneous mutations in the creation of proposed utopian city spaces. In fact what they propose at the Musee d'Art Moderne is the artificial growing of extruded urban housing (generative & robotic) where new cities are constructed via robotic processes[328] by feeding off the carcasses of older dying cities. Very noisy-viral. Envisioned is an approach to city planning based on growth scripts and open algorithmic procedures.

Towards these ends, the show itself includes some subtle audio tracts, model-sculptures, a fully immersive hypnosis chamber with video monitors, booking services, 3D movies and robotic drawings/plans that reveal the source code of the generative program at the heart of their work. There is a definite tangled and intertwined approach to the city vector that reminds me of the dithyrambic noisy visual hyper-logic which has manifested in all modes of decadent artistic periods. The multiplicity of its interwoven experiences challenges the now bogus idea of simplicity, a modernist-minimalist idea that has taken on the intensity of a righteous injunction in many cases. Given the organic-like, biomorphic architectural forms R&Sie(n) spawned by their generative program, I could not avoid thinking about the visionary city-planning put forth by the Situationists. One thinks immediately of Guy Debord's essay *On Wild Architecture*, for example. Like the Situationists, for R&Sie(n), the urban form no longer depends on the arbitrary decisions or control over its emergence exercised by the elite few. Ultimately R&Sie(n) lead me towards juicy Situationist-like complexities and engagements by way of immersion into an open-ended noise field.

As R&Sie(n) say themselves, "Many different stimuli have contributed to the emergence of *I've heard about...* © and they are continually reloaded. Its existence is inextricably linked to the end of the grand narratives, the objective recognition of climatic changes, a suspicion of all morality (even ecological), to the vibration of social phenomena and the urgent need to renew the democratic mechanisms. Fiction is its reality principle…". Ah, the domain of decadent noise artifice.

What has been somewhat poorly determined however is the degree to which a dweller feels totally immersed in an optically excessive space. And this depends to a large extent on personal psychological need and adaptability in accord with the proposed spatial depth cues. Cognitive-aesthetic space has to be coordinated phenomenologically with the proprioceptive space of the eye—and R&Sie(n)'s only failure is in maintaining the evident structural seams of the immersive faux-hypnotic chamber (the only enterable structure and the highlight of the show) because what the entire show is proposing is a seamless immersion into generative noise, and the visual seams take us out of that exquisite fantasy. So they are denied the loveliest of triumphs.

A small pity, as one might otherwise imagine oneself totally immersed there somewhat like a 21st century dandy. Like those at the birth of the 20th century, this new hyper-dandy constantly might affirm his or her originality down to the decorative details of the home. In that the robots are doing the algorithmic planning and building, this work unquestionably proposes a new form of dandyism—if dandyism's defining characteristic is remembered to be the making of one's person a work of art while extolling laziness and displaying contempt for work. Evident here are the Baudelairean/Duchampian dandy ideals of impassivity, nonchalance elegance, and inscrutability. What matters are the triumphs of a radical contempt for one's hand.

Indeed, one can say that *I've heard about…* © favorably extolled artificiality, indifference, impassiveness—the reign of an ironic causality and knotted ambivalence—while staying open to all noise transactions. Most importantly, a-life forms are embedded within it and its growth is artificial and synthetic. So R&Sie(n) maintains a version of transcendental phenomenological idealism, but they do not disavow the extant actuality of the material sphere. Instead they seek to elucidate the sense of the world-as-is today—that is viractualized—by stressing the embodied nature of human and artificial consciousness and bodily existence as the original and originating material premise of sense and signification.

CONCLUSION

Noise Against Oblivion: An Omnijective Philosophy of Noise Culture

To think about existential problems in such a way as to leave out the passion is tantamount to not thinking about them at all, since it is to forget the point.
— Søren Kierkegaard, *Concluding Unscientific Postscript*

I now feel capable of evaluating the evidence. There is for me an evidence in the realm of flesh which has nothing to do with the evidence of reason.
— Antonin Artaud, *Manifesto In Clear Language*

Philosophical theory is a practice of concepts, and it must be judged in the light of the other practices with which it interferes.
— Gilles Deleuze, *Cinema 2*

In agreement with the philosopher Richard Rorty's *Philosophy and the Mirror of Nature*, my expectation in espousing this entangled immersive art noise theory is that it will be understood not as a static aesthetic position, but as a glitch move that is continuing to amplify. This theoretical glitch hypothesis makes no pretensions to scientific methodology as it, by its very nature, must retain its speculative character because of the impossibility of attaining conclusive experimental data in art. A similar impossibility is also the case, however, with the field of consciousness studies,[329] a field which often insists on calling itself scientific. So, rather than contending that this book's conclusions are scientific, this investigation into cultures of noise ends by involving its findings in the artistic phenomenological question of the *qualities (and levels) of awareness of our own consciousness* within aesthetic realms which we are capable of attaining through noise art.

Especially inasmuch as I am involved in the humanities, I am reluctant to model my conclusive methodology on a mechanistic model of an earlier power-oriented science, even though the philosopher Werner Heisenberg maintained that the differences between art and science are minimized if one views both art and science from the more general vantage point of the Zeitgeist. Indeed, Gilles Deleuze also points out that "the special perceptions and affections of science or philosophy connect up with the precepts and affects of art".[330] Therefore, in that "conscious experience is not directly observable in an experimental context",[331] it is indeed this Deleuzian science/philosophy/art connected phenomenological zone that seems to be the appropriate theoretical model for this art noise summation. The philosophic rhizomatic theory[332] of Deleuze and Guattari, at a general level, supports such a connectivist-glitch approach towards theorizing art, as rhizomatic theory encourages philosophic non-linear and non-restrictive thinking/imagining.

For me, Gilles Deleuze's vision of our post-industrial life opened the way for the production of current noise art by affirming the befittingness of multiplicity and the necessary entitlement of dissension. In recognition of his work, this reflection has been an attempt to hypothesize and demonstrate a counter-mannerist glitch excess, an idea that was specifically inspired by the rhizomatic thinking of Deleuze and Guattari. This glitch idea re-establishes an ambiguously private critical distance for art: a distance achieved through the connectivist challenge of (and disparity between) pleasure and frustration. This theory of noise art demands of society an *active visualizing participation in private interpretations*—and thus is a legitimate metaphor for contemporary art as a form of simulation-shattering engagement. For example, the noise music of John Cage demanded an open mind from the listener and a predisposed ear for all the sounds usually excluded from music in the traditional sense. If (s)he, the listener, can relate to the "noise" sounds (sounds that are always already there anyway) in a "musical" sense, then the distinction between music and noise becomes very diffuse, tentative, and rather arbitrary.

This reminds me of Raymond Roussel's themes and procedures that involved imprisonment and liberation, exoticism, cryptograms and torture by language—all formally reflected in his working technique with their inextricable play of double images, repetitions, and impediments, all giving the impression of the pen running on by itself through the dreamy usage and baroque play of mirrored form. Roussel's technique and the process he developed lends itself well to the creation of unforeseen, automatic and spontaneously noise art which gives me the feeling of prolonging action into eternity through the ceaseless, fantastic constructions of the work itself, transmitting an altered, exalted and orgasmic state of mind which after the initial dazzling creates one predominant overall effect, that of creating doubt through mechanical discourse.

The image of enclosure is common with Roussel where a secret to a secret is held back, systematically imposing a formless anxiety in the reader through the labyrinthine extensions and doublings, disguises and duplications of his texts, which make all speech and vision undergo a moment of annihilation. Roussel presents to us the model of noisy perfection of the eternally repetitive mechanical machine which functions independently of time and space, pulling the artist into a logic of the infinite.

Roussel's last book, *How I Wrote Certain of My Books*, is the last of his conceptual noise machines, the machine which contains and repeats within its mechanism all those mental machines he had formerly described and put into motion, making evident the machine which produced all of his machines—the master machine. All of these machines map out a noise space that is circular in nature and thus an abstract attempt at eliminating time. They reproduce the old myths of departure, of loss and of return. They construct a crisscrossed mechanical map of the two great mythic spaces so often explored by western imagination: space that is rigid and forbidden, containing the quest, the return and the treasure (for example the geography of the Argonauts and the labyrinth)—and the other space of polymorphosis noise, the visible transformation of instantly crossed frontiers and borders, of strange affiliations, of spells, and of symbolic replacements (the space of the Minotaur).

With such a wildly visionary[333] look at an art of noise, I wish to suggest what art's contribution could be to the enlargement of self-understanding[334] in the context of our conspicuously excessive, connected and collapsing society. My contention is that glitch awareness/appreciation potentially removes us out of our quiet and glib indolence and points us in the potent direction of expanding thunderous intensity.

I believe that a post-Pop noise art is critical to us now[335] because its glitch counter-mannerist excess can problematize the popular simulacra and make livelier the underground privateness of the human condition while remaining immersed inside the social network that engulfs and (supposedly) controls us. This glitch consideration offers us a personal critical distance (by skip, by stutter, by gap), and thus another perspective on (and from) the given social simulacra.

Such a destructive-creative thought might provide us with two essential aspects relevant to our lives. First, it can provide a private context in which to suitably understand our current situation.[336] Secondly (but more importantly), it may then undermine this understanding by overwhelming our immersion in the customary— along with our own prudent pose as judge.

For me, then, a bacchanalian post-Pop glitch art is capable of functioning (paradoxically) by nurturing in us a sense of polysemic uniqueness and of individuality brought about through a counter-mannerist destructive-creative style (ever more circuitous, excessive and décadent)—a style that takes us from the state of the social to the state of the secret I, by overloading ideological demonstration to a point where it becomes non-representational. This destructive-creative thought makes judicious use of the process of Deleuzian/Guattarian nomadic thinking (hearing/seeing).[337]

Accordingly, Deleuzian/Guattarian noise art would be composed of variously formed segments, stratas, and lines of flight which involve territorializing as well as deterritorializing spacio/psychic activities.[338] It is this nomadic, non-representational, counter-mannerist vacuole glitch that can suggest breaks from the fascination and complicity with Pop art and the mass media mode of communication (hung up on the felt need to be *liked*).

As I have shown, the art of noise needs not be likeable—nor be polite. It is certainly not info-tainment. It is, rather, as Paul Hegarty points out, *infliction*.[339] For me it is, more precisely, an *infliction of a pleasant frustration that can lead to creative visualization*.

Subsequently, my idea of noise art can also be long-suffering *drôle merde* (funny shit).[340] Undeniably, such a comic art backflip ties into the counter-mannerist pataphysical anti-concepts developed by Alfred Jarry (1873-1907) and The Collège de Pataphysique founded on May 11th, 1948 by an anarchic group of artists and writers interested in the philosophy of *Pataphysics*. These zealots devoted their time to perpetuating (and often distorting) Jarry's philosophical pranks.

For Jarry, Pataphysics is the anti-scientific realm beyond metaphysics that examines the laws that preside over *exceptions*—an attempt to elucidate an imaginary cosmos. Jarry specifically defined Pataphysics as the "science of imaginary solutions, which symbolically attributes the properties of objects, described by their virtuality, to their lineaments".[341]

In 1959, Marcel Duchamp agreed to be a *satrap* in the Collège de Pataphysique, and there have been numerous links established with the Oulipo literary movement, specifically through the participation in both groups by the poet Raymond Queneau (1903-1976). The fabulous wordsmith, Jean Genet (1910-1986), has described himself as following in the Pataphysical tradition. Pataphysics makes a rare pop musical appraearnce in 1969 in the Beatles' song "Maxwell's Silver Hammer" on *Abbey Road* by mentioning a student who studied pataphysical science in the home. But unlike pop culture as a whole, Pataphysics (like noise culture) navigates inside popular topographies by disturbing their sense of collective order.

Like the thinking of Jarry, art noise mostly arrives at an anti-social examination without demonstrating any sustained systematic analysis. It just interrupts. That is what makes it counter-mannerist. But what strikes me as most exact to noise art's peculiar propositions in terms of Pataphysics is its deep reflection (one might even say brooding) on the theme of ignobility, and this realization should shoddily shift something in your appreciation of it. Notably, already evident in the art of noise is its display of a harsh obsession with volume, a volume (both visual and audio) that tests the limits of form and stretches the bounds of meaning by recasting our experiences of encountering wildly disjunctive ideas into the sumptuously physicality of negation.

This reality-rejecting-ness delivers an airy irrational punch of nonsensical negation to art noise by tying together methods of insouciant informality with a visceral irony: at turns hip and flamboyant or abrasively outrageous. Still, the audience is expected to work devotedly to appreciate its absurd conundrums, to supply mental transitions between the diverse assortments of irrational elements that supply the art of noise its Pataphysical hooks. One must fabricate a complicated forensic fairy-tale out of this counter-mannerist mélange as it keeps slipping in and out of idiosyncratic narration. And, alas, that recitation keeps turning back into one about stinking death, that strange, incurable and deeply irrational affliction.

So yes, as noise art is about self-transcendence by means of rupture, I read the art of noise as a meditation on humiliating death in all its undifferentiated fabulousness, by which I mean its essentially nasty comedy. It is a counter-mannerist art about comical, difficult death, then. Pulling down our pants and revealing our soiled undies while keeping everyone laughing (or at least gurgling) till the deafening end.

With the art of noise, there is then an awareness of impertinent splendor in the tranquility of decomposition, which makes it all seem faintly heroic in face of death's inexorability. Thus this irrational art implies an antiphilosopher's knowledge of dumb death's putrid ignobility—but the art of noise will not give in to that parody either. And this is what gives the work its extraordinary sense of dignity, a dignity that asserts life's primacy over death because death is beyond images, beyond sounds and beyond words.

Accordingly, art noise's hypothesis is actually fine absurdist Ubu art. Ubu is first encountered in Alfred Jarry's *Ubu Roi*, a play that created a famous scandal when it was first performed at the Theater de l'Oeuvre in Paris in 1896. It is an important precursor of Dada. Through a language of shocking lad hilarity, Ubu Roi tells the farcical story of Père Ubu, an officer of the King of Poland, a grotesque figure who epitomizes the mediocrity and idiocy of middle-class officialdom. It was through writing Ubu Roi that Jarry became the creator of the science of Pataphysics, his absurd a-logic that defined the science of imaginary solutions as enshrined since 1948 in the Collège de Pataphysique. But an Ubu art does not merely help us pass the time away, it enlivens time if we surrender to its fearful difficulty—as noise art may provide the chance to do the counter-silent thing, to look at and hear what we fear, so that such an effort will help release us from fear's irrational grip. Then we might pataphysically expand into the counter experiences of noise and see beneath the stucco surface of *Maya*[342] and so enjoy absurd life all the more. So that the ignobility of death can be ignored and nonsensical dignity restored—for the fleeting moment, at least.

Deleuze and Guattari's term for such counter-experiences to mannerism[343] is *becoming-animal*. For them to "become animal is to participate in movement, to stake out the path of escape in all its positivity, to cross a threshold, to reach a continuum of intensities where all forms come undone, as do all the significations, signifiers, and signifieds, to the benefit of an unformed matter of deterritorialized flux, of nonsignifying signs".[344]

Building upon their suppositions, I speculate that glitch excess can put forth an aesthetic élan of superabundance that re-conceptualizes art in terms of noise connectivity so as to unbridle us (some). This is how I interpret the counter-experience of feeling intricated in a becoming-animal noise panorama by fashioning a map of intensities.[345] However, this glitch character of de-simulated openness, which an inception of the art of noise assumes, demands that we seek a liberation from custom, doctrine and influence, and that we grasp again the autonomy and priority of art as a special type of excessive ideological activity.[346]

According to Deleuze and Guattari, rhizomatic activity is boundless in its branching. Thus noise art reflections may cross wide chasms of psycho/optic space on one surface as the most disparate elements and details may be linked. Moreover, a psychic rhizome is continually dynamic and is ceaselessly actualized by the arousal its dynamism produces and thus it is never in accord with some pre-established strategy or imposed configuration. The psychic rhizome is regularly swarming itself into being as micro and macro factors attract. One cannot declare in advance what its limiting confines are or where it will or will not operate nor what may become connected and tangled up in the rhizome's multiple dimensions because the connections do not inevitably plait common types together. Rather, a psychic rhizome's multiple dimensions instigate cross-overs between both the highest synthetic level and the slightest, most minute discrete distinctions. An artistic noise rhizome would be a complication of perceptual vicissitudes so intertwined that it gives birth to different scopes of macro-perception.

Such a noisy probing at the outer limits of recognizable representation and the excited all-over fervor of such a syncretistic probe isn't a failing of communications within noise art terms then. *It is its subject.* Such a bountiful realization is insinuated through overloaded/excessive stimulus inasmuch as noise may represent every integrated meaning conceivable (as white noise) for, as I demonstrated, in the art of noise the focal/audio point is generally uncircumscripted. The expansive elements within noise art are not, by definition, passively received and accepted then.

By nature of its challenging, conflicting, excessive presentation, visual noise art is also psychically withheld as it is inexorably displayed all at once to the limited nature of our human perceptive competence. Thus a successful art of noise takes us away from the habitual focus of the picturesque and potentially liberates us inwardly from the infringements stemming from the deluge of mass-media images—and so stimulates us to assess anew the caliber of any such infringement. Hence it is in the amity felt with the glitch that we may feel a sensuous liberation from ideological monotony and cultural prudery.

Such an art of noise may then stand in defiance against the limits of ordinary perception and representation. Thus it is (or can be) about the opposition between the daily work-day and the transgressive/ecstatic moment. In a sense, it attempts to set up a long-standing form of ecstatic offence where one can go back and forth at will via the glitch.[347]

Underlying this aim is a miasmatic idea which questions linear and hierarchical structures and seeks to replace them with atmospheric loose structures, keyed to a penetrable, reciprocal flow of events. This realization came to me when reading Gilles Deleuze's *Spinoza: Practical Philosophy*. In it, Deleuze pointed me towards a recognition of my desires' productiveness, as he indicated how desires propel us to move towards greater or lesser states of magnificence "depending on whether the thing encountered enters into composition with us, or on the contrary tends to decompose us...".[348]

So as noise art both decomposes us and stirs re-composures in us, this means a re-positioning of identity within an atmospheric and artistic ontological model of relations, diversities, shocks, connections, heterogeneities, breaks and unexpected links. This means that the intellectual situation of Deleuzian inspired noise art is one of magnanimous self-connectivity through confrontation.

As such, noise art promotes *as if* imaginable desires for extrasensory[349] distributed disembodiment[350] involving the enthused transmigrational expansion of boundaries and a yearning to penetrate and merge characteristic of spirituality (*ignudo spirto*). Hence the role of noise art remains the necessary prosthetic task of artificially facilitating such an unrestricted state—and so remains associated with the most fleeting elaborations of artistic consciousness. This erudite desire to exist in an anti-mechanistic state of expansion is temporarily realized (albeit symbolically) in engaging noise art. As such, noise art posits itself as a meta-symbol of and for expanded human potential linked to tolerance.[351]

This goal of an expanded human capability through art is important to me, as I feel that the substantial ability to self-modify (self-re-program) ourselves is the point of art. In this inference, aesthetic immersion into noise adheres to and fosters Kendall Walton's *theory of make-believe* in which Walton sees art as a generator of *fictional truths*[352] which through art's inventiveness invites ontological self-modification via participation in the creative process. Moreover, Walton's theory of fictional truths reflects Friedrich Nietzsche's important assertion that "logical fictions", which he saw as "comparisons of reality with a purely imagined world of the absolute", are indispensable to humanity.[353] The key value of immersive noise's fictional truths in terms of formulating an original theory of noise art, however, lies in underscoring the fiction behind the assumed "real" perspective[354] when seen as "empirically true and universally valid" instead of as "conventional and contingent"[355] idiosyncratic compliances.

Given that noise disturbs order, it is reasonable to interject here that the notions and experiences of aesthetically quickened disembodiment may (via noise art) claim the distinction of serving as the (or a) lucent interface between David Bohm's aforementioned implicate order and explicate order. But aesthetic noise consciousness above all renders a *lightness of being* which is supported by a *metaphorical consciousness of passage* principled on the electron transport conditions of the nerve cells. Hence aesthetic glitch sensibility is rooted in linked neurological self-programmable operations where the conceptual exchange between the disembodied/ecstatic and the bound/submissive (conceived of as teeming), constructs the neural-noise-metaphysics[356] of immersion into noise—as well as, what Vladimir Nabokov (1899-1977) would consider to be, its "combinational delight".[357]

As I have outlined earlier, fixed-point perspective generally configured pictorial art in the West since the Renaissance. Immersion into noise's fundamentally spherical, all-over perspective of dynamic thresholds cast a fraction of art on its course since the Fin-de-Siècle. This marginal tendency has now amply flowered in noise art as art practice began shifting us away from illusionistic *trompe l'oeil*. Immersive noise-art's 360° cognition enlivens receptive and organizing attributes of peripheral awareness and, as such, intensifies thalamic input to the cortex by making the active thalamic neurons in that region fire more rapidly than usual. Moreover, with this immersive noise vision there is a shift to a more conscious peripheral mode of perception which entails a de-automatization of the perceptual process (whereby more emphasis is placed on what is on the edges of sight, sound and consciousness) thus presumably adjusting us to an expanded and fuller consciousness. This emphasis on the peripheral utilizes the Deleuzian *broad scan*, Deleuze's non-linear dynamic conceptual displacement of a view along any axis or direction in favor of a sweeping processes in space/time.[358] Hence, immersive noise vision may acquire an increasingly computational-like encompassing range useful in expanding the customary perception so as to increase situational awareness. For, as Luigi Russolo said in his seminal text *The Art of Noises*, "Every manifestation of our life is accompanied by noise".

Noise consciousness is essentially a cognitive challenge to our habitual sensibility, then—a challenge to find the fullest possible cognitive resources to cope with the expanded context of the art's excess, and a proposal that implies that those cognitive-visual resources are available, if as yet undefined. Hence aesthetic immersion into noise is consistent with Georges Bataille's intellectual comprehension of dithyrambian excess (itself suggestive of the human cortex with its vast array of micro intra-cortical nerve connections) as a mercurial movement that surpasses entrenched limits. The *intensity* of indeterminate dithyrambian excess as experienced in dynamic noise art is key to this cognition.

By refusing the dichotomized, utilitarian, manageable codes of representation with free non-logocentric associational operations, noise art triggers a multitudinous array of synaptic charges and thickens perception to the extent that it prevents the achievement of a prior determinate aesthetic. This threshold component of the immersive noise aesthetic adds enough uncertainty to the usual signals in the internal circuitry of the human biocomputer so as to make new configurations of the self probable (by organizing the internal energies of the self more broadly via disembodiment). *The subsequent and ultimate aesthetic benefit of noise art, then, is in attaining a prospective realization of our perceptual circuitry as a self-re-programmable operation.*

This self-re-programmable ontological operation occurs specifically in a constructed space between the noise art and the subject, similar to how Wolfgang Iser locates the encounter with a written text by its reader in a third realm of indeterminate interaction which he calls the *work:* a transaction situated "somewhere between" the text and the reader.[359] However, unlike a written text, the self-re-programmable ontological adjustments and modifications one makes during the process of coming to understand noise as art never ceases and sensorial closure is never evoked.[360] This kind of liberational ecstasy potentially found within aesthetic noise is exceptionally important, for as Brian Massumi tells us, "if there were no escape, no excess or remainder, no fade-out to infinity, the universe would be without potential, pure entropy, death".[361]

Aesthetic immersion into noise's indeterminate vibratory excess, then, facilitates our desire to transcend the boundaries of our customary human cognition so as to feel that state of unconditionally Hegel called the *absolute* (our absolute sense as an unalloyed being akin to non-being) by way of a neuro-metaphysics conveyed through noise art's ambient milieu. This vibratory extrasensory dispersion, which presupposes a loss of fixed reference points, implies a diaphanous neural-metaphysics constructed around the disembodied psyche's enhanced identity as *non-site* consistent with Jean-François Lyotard's (1924-1998) assertion that metaphysical concepts have been realized in the contemporary world.[362] By the noisy psyche taking up an anti-position of circuitous non-site, I ascertain that the immersive noise sensibility is essentially *non-logocentric, ecstatic, variational, non-hierarchical,* and *excessive*. It is particularly *excessive* in that immersion into noise deframes and overwhelms the envelope of hardened fixed-point (i.e., window) perception with aesthetic input and, hence, is an excess of and for the prosaic gaze as it offers an immersive scope beyond our perceptual capabilities. Indeed, instead of nicely proceeding along towards an expedient comprehension and appraisal, immersive art noise actually opens up an oppositional anti-mechanistic space of self adumbration for the self-re-programming, ontologically-minded by revealing the loose limits of our solipsistic (and hedonistic) inner circuitry. The noisy excess necessary for triggering such an immersive keenness offers to the self-re-programming immersant a scope of sensibility beyond what Jacques Derrida identified as typical of the consolidated, passive, spectator/consumer.[363] Indeed, in our heavily materially oriented, technologically accelerated, information saturated culture, where experience is increasingly prescribed, facile, and fast, *thoughtful langor in noise* satisfies an essential need consistent with the interpretative theories of both hermeneutics and the phenomenology of perception. In this respect, noise fulfils the *negative dialectical ideal of art* as affirmed by Theodor Adorno when he upheld the view that the radical potential of art lies in formal innovations which refuse to allow its passive consumption, demanding instead an active-critical intellectual involvement (inherent in the noise aesthetic) in opposition to unthinking assimilation. It is for this reason that noise art possesses a negative dialectical felicity of its own. For as Luigi Russolo says in his seminal text, *The Art of Noises*, "to attune noises does not mean to detract from all their irregular movements and vibrations in time and intensity, but rather to give gradation and tone to the most strongly predominant of these vibrations".

The negative dialectical confrontation with non-knowing typical of noise is an important component of noise-art-consciousness' intellectual satisfaction, as the entire benefit of addressing the espoused principle inherent in noise art exists in attempting to adhere to an exciting transmissible hyper-state that exceeds, ruptures, transcends, and overwhelms our former inner territory.[364]

Noise culture (wired with an excess that consequently disallows itself to be readily grasped) suggests a sense of *immense inner incompleteness* commensurable with a post-Hegelian consciousness, as Hegel maintained that no idea has a fixed meaning and that no form of understanding has an unchanging validity. Indeed, this post-Hegelian consciousness of excess is how immersion into noise challenges distinctive ontological beliefs about the limits of the self. In immersion into noise, self-re-programmable thought takes over the space displayed around the constructed self as the meta-programming ego expands to fill (by transference) the vastness of noise. So conceived, the ontological self ceases to think of itself as a substance or thing and, by contrast, thinks of itself as a continuously changing process of vibrational events.

This would indicate that, when challenged with art noise, our bio-circuitry—circuitry usually occupied with perceptual demands of a different (more standard) kind—must surpass our prior limitations (and lazy assumptions) as programmed by the petite bourgeois world-view. This self-programming variance, however, requires (and is) what Søren Kierkegaard (1813-1855) calls *passion*. Indeed, it is in this manner that I wish to use the term *hyper-cognitive* with respect to noise art, a term that has been extended out from Dick Higgins's term *post-cognitive* which he put forward to describe the field of post-1958 artistic production centered around developments in *Intermedia*[365] and Happenings that conceptually fused together aspects of traditional media so as to promote non-linear and multi-linear thinking.[366] Higgins adapted the term from the Fluxus related artist/non-artist/philosopher/noise-musician Henry Flynt's use of *cognitive*, a term that addresses the issue of becoming through connectionist activity which Flynt approached from a post-Logical-Positivist position he called *beliefless empiricism*.[367] Flynt's 1961 text "Concept Art" (first published in book form by La Monte Young and Jackson Mac Low in the 1963 publication, *An Anthology of Chance Operations*) outlined the genre that later became known as Conceptual Art and had a strong impact on me. According to Flynt, Conceptual Art is "an art of which the material is *concepts*", whereas Higgins asserted that "the process of cognition", after 1958, was "no longer a major subject for artistic exploration".[368]

I find that, with the necessity of understanding noise culture, cognition (now viral-cognition) is again a dominant theme for art. Indeed, I believe I have shown here that an allusive sense of embedded viral-cognition of productive paradox is indeed the inner logic (and one might also say the poetics) of the art of noise.[369] This finding collaborates what Siegfried Zielinski thinks is the potentiality of virtuality as an aesthetic enterprise in that Zielinski maintains that art's potential in terms of virtuality lies in the "tension between the virtual and the impossible".[370]

Viral noise culture expands the measure of cognitive-perception through well-being's enthusiastic contact with impossible (and disturbing) excess and, thus, is a superabundance of and for the imagination. Insofar as ontological distinctions within noise culture clatter, it makes extensive demands on our previous facilities of critical reflection by addressing our inarticulate connections between the intuitive and the conscious realm that cannot easily be expressed in words but of which we can be occasionally appreciative.

Cultural noise, then, is identifiable (and interesting) when it sounds or appears as moments of non-narrative narrative—where something runs counter to the narrative imperative of the human psyche. Thus, fully understanding the non-narrative narrative of noise culture is only possible through addressing the question of consciousness called *cognitive dissonance*. Cognitive dissonance is a psychological term denoting the mental state in which two or more incompatible or contradictory ideas are held to be equally sustainable. A person who is successful at keeping contradictory ideas in dialectical suspension is said to have a high degree of *negative capability*. This is a particularly important concept for noise culture for, as Deleuze asserts, there is a "double structure to every event".[371] Also, we must recollect once more Massumi's Deleuzian interpretation of the virtual as "a lived paradox where what are normally opposites coexist, coalesce, and connect".[372]

So yes, *Noise Culture* as a perturbing designate of non-narrative narative may appear oxymoronic and hence cognitively dissonant. Subsequently, such a vacuole theory of paradox would be aligned with Gene Youngblood's theory of synaesthetic cinema, an alloy, superimpositional cinema[373] which creates an "awareness of the process of one's own perception",[374] by being structured by the language of paradox.[375]

From a sociological point of view,[376] noise culture opposes (through enhancing our awareness of the plurality of the possible) existential postmodernistic oblivion—the dulled, blasé, over-stimulated gaze of the alienated and ironic. It does so in that the practice of noise-culture-mindfulness requires continual discernment with the mind in an *imaginative* and *receptive*, rather than passive, state. Hence, noise culture may provide a fundamental antithesis to the authoritarian, mechanical, simulated rigidities of the technical world. Indeed, aesthetic immersion into noise stands in opposition to mental (and cultural) sleep in that its methodology is to jolt, exceed and hence attempt to expand with appreciative input. Correspondingly, ontological consciousness expands into an enlarged synthetic field of vibration: the connective field of *inter-consciousness*. In this field-connective-condition, notions of a singular, discrete, logocentric consciousness are incoherent.

In this respect, Charles Baudelaire's ode *Paradis Artificiel* (*Artificial Paradise*) is increasingly outstanding for what it has to say about the promise of noise culture. In it, Baudelaire cites imagination as that force that serves noise culture's primary purpose, which is to *enlarge consciousness so that it may approach the connective field of inter-consciousness*. Georges Bataille confirms this assertion in his essay *Baudelaire*, particularly by linking Baudelaire's imagination with notions of the impossible. As noise culture places us in the position of indeterminate unknowing, consciousness ceases to be definitive and becomes inter-relational, and so provides questions that disturb and disable previous emphasis on the false objectivisms accorded to cultural production.

This extremely pliable state of cognition rests on the basis of relaxed (but extended) attentiveness. Indeed, it is within this elastic attentiveness (constituted between artist and audience) that noise culture takes place. In this respect, Giovanni Careri's description of Bernini's *bel composto* as a "vehicle of an experience that goes beyond all images"[377] pertains to the central idea of noise culture. By way of dishabituation and disorientation through merging with the *mise en scène* of noise, a partial obliteration of self-consciousness may occur which returns self-representational mimesis to the infinity of noise in the raw sense. In that sense, immersion into noise is a tender *via negativa*, a delicate overrunning and self-annihilation in the interests of further growth and expansion.

That said, I contend that what is also important about noise culture is that which is revealed as being in synthetic accordance with the basic extravagant and excessive aspirations of humanity, in accordance with the effective deployment of the earth's sustainable resources. Such a notion of responsible excess is important to the extent that it helps us push deeper into our stored ontological metaprograms, so as to enhance their (our) performance in the social and ecological field. As such, noise culture is an exercise in conscious responsibility coupled with liberty produced in the reflective province which Baudelaire calls *art*.

Thus, my idea of noise art adheres to Antonin Artaud's proposal in *Le Théâtre et Son Double* (*The Theater and its Double*) that art (in his case drama) must be a means of influencing the human organism and directly altering consciousness by engaging the audience ritualistically. Even though in his essay, "The Theater of Cruelty and the Closure of Representation", Jacques Derrida describes how Artaud's theory may be seen as impossible in terms of the established structure of Western thought,[378] this is precisely why noise cultural theory (with its, as previously explained, vital connections to the unbearable) can be placed in parallel position to Artaud's hypothesis. This is so in that, when inside noise art, one experiences a prelude to the work's fullness (its impossible commotion usually diverts the immediacy of the art), thus stimulating a viral desire that bio-chemically affects the state of mind.

In noise culture, we are essentially challenged to find new expanded boundaries.[379] For example, John Cage does not depart from music when he begins with noise. More accurately, he creates a music that belongs to the noises of the environment and takes them into consideration. As in Artaud's theory of cruelty, this challenge to find new expanded boundaries through noise is accompanied by a vibrating love-hate where we may *sense an intimate—if disagreeable—relationship to the work*. Through the immoderate excess of perceptual possibilities in an aesthetic immersion into noise (which involves a more active and continuously searching situation) one enters (and thus hopefully understands through experience) hyper-noise, a state which circumvents the current fragmentary view of the self in the world that has been built into the structures underpinning traditional culture.

I believe that my noise art examples have amply testified to the potential for noise culture—and, by concentrating on myself in relationship to noise art, I have articulated intense (but short-lived) art noise experiences that will not let me reach a final organization. Subsequently, noise forms a semi-living viral[380] culture where the form and the meaning construct a precise union in their agitational existence. This precise union suggests a taste for Mediterranean paganism's stoic view of life where the arbitrary cruelties and difficulties of being are met without flinching. Here, forms of extended vitality are cherished and diligent forms of stasis shunned, including dreary scholarship and most aspects of routine public life—so as we may respond to all forms of art more vividly and completely, not merely the thwarted or offensive. So noise sophistication is about developing our quality of disinhibited mind as manifested in full sensuous reactions.

The primal wantonness suggested in edgy art noise sophistication epitomizes what is exhilarating about culture. It rips open our delirious heart of darkness when noise (typically thought of as a bad quality) and art (typically thought of as a good quality) mingle. Hence hostility and anxiety mix with what lushly attracts us. And this is the magic formula for excitement: obstacle mixed with attraction equals excitement.

As we have seen, the scintillating genesis of noise culture starts in the semi-artificiality of the adorned prehistoric cave and points towards digital interconnection and viral rupture. But the lax digital (interconnected) signal fragment, and the speed and ease in which it moves of late, has now lulled our emotions and is beginning to fail to stimulate critical thought, transformation or perturbation. What I deduce from this history is that noise culture is brimming with hypotheticals, when adversary. The noise culture theory put forth here then clarifies, deepens, confirms and exalts a desire for elite, virtuoso, and indeterminate art noise that, as I have shown, has marginally existed in human culture throughout time.

This stoic theory of viral-noise culture is one of vibrational estrangement: breaks and bindings that admit unknowingness and the non-self. Its high seriousness insists on difficulty and is a reinvigoration of vangaurdism. As such, it is qualitatively and quantitatively distinct. Its politics are those of listening hard and seeing widely. Its *esprit de corps* is diaphanous being within a kind of experiential span and ocular extravagance through which abstract inter-relationals are felt to be a function of one another. For when seeing our well-being connected to states of exuberant, non-graspable noise, we temporarily dissolve former boundary ontological dichotomies into an unoccupied topos of being that implicates our internal self-programmer in a desire for unrealizable summational resolution.[381]

Noise art thus serves art's classical function of modelling in microscopic proportions the overwhelming phenomenon which Hegel and Nietzsche called the *absolute* (an imagined orb beyond immediate sense perception). As such it addresses, by extrapolation, self, social, and political limitations when we adapt and incorporate aesthetic noise (with its insinuation of natural infinity and its inexorable pull towards the liberation of confines) into our everyday view of the world.

In the space of linear perspective (with its historically invented fixed placement of the vanishing point on the horizon), the viewer imagined that she or he was looking at the complex world as if through a window—as a spectator. This implied a separating wall with a small rectangular opening in it between the subject's perceptions and the visible world as dictated by pre-established criteria. This pre-established window has become the post-industrial human's mental tendency. Behind this pre-established window, human sensitivity has become increasingly bored, neutral, distant, detached, separated and narrowed. With this tendency, restrictions on the excess of complexity are already decided unconsciously in advance, regardless of the actual character of our perceptual input. As a result, there has been a de-emphasis on the peripheral, and the ambient as vision has become restrained by the habits of linear perspective; pre-established habits now encoded in the methods and expectations of photography, video[382] and film.[383] Thus vision has increasingly taken on the attributes of a focused, singular, narrow vision which is staring straight ahead.

In contrast, noise culture proposes a style of consciousness marked by an emphasis on din and by a re-entry into the rich fringes of sensation. Keep in mind that John Cage's music starts from the simple fact that we are always already surrounded by noise. What is vital to our consciousness is how we connect to those noises. Cage suggests a lucid scheme: if we try to disregard them, they agitate us; but if we listen to them and recognize them, they may become enthrallingly artistic.

The viral noise theory of consciousness that I have been sketching out here is not a precise theory of consciousness in that it does not explain what consciousness is, nor does it explain exactly how it arises in the first place. Such problems of defining the essential constitution of consciousness have been widely discussed elsewhere within the realm, principally, of philosophy, thus far without arriving at a consolidated consensus. But if we accept the more modest definition of a theory of consciousness as a theory of self-awareness of how our inner life and thoughts function (and may function fully), I take it that what I propose concerning noise consciousness might be judiciously placed within the arena of contending theories of how consciousness functions (and/or may function)—in this case, through ruptured induced expansion.

Lateral (horizontal) thinking, a term introduced by Edward de Bono, refers to the capacity to shift the context of thought away from conventional logical (vertical) progressions to unaccustomed lateral ones, thereby shifting thought away from fixed, predefined orders and towards creative ambiguity. According to Albert Rothenberg, "creation involves intense motivation, transcendence of time and space, and the unearthing of unconscious material",[384] and lateral thinking is beneficial for shifting consciousness out of habitual formulations and "ways of seeing".[385] In lieu of my ideas of intense noise culture, creative immersive noise thinking might now be conceptualized not as a vertical or even lateral thought process, but as diaphanous and spherical. Such a boisterous and diaphanous formulation defines noise culture's general pull away from established thought and is what makes it, in Paul Hegarty's words, a form of *anti-fascist*[386] *resistance*.[387]

Noise art, with its implied access to the ineffable, suggests gentle chaos, which is of course the basis of our most advanced revolution in scientific understanding: chaos theory. Still, the art of noise, for the connoisseur, is generally a ribald art of the outside(r) where noise culture feeds into a cognitive process that involves a deep involvement in (and appreciation of) the contradictory nature of opposites and antitheses (now blended into living abstractions). Such creative thought is useful in configuring a viral-oriented cultural vision of the technological world sensitive to what John Cage made clear: that all the music we hear—and I add much of the art we see—is constantly and inevitably pervaded by noise, by uninvited vibrations. In this operationally defined model of the creative intellect, artistic and divergent thinking wins the capacity—through situating itself within an immersion into noise—to generate and appreciate multiple alternatives by deviating away from overly hushed modes of perception-cognition. Such sensitivity is enhanced by experiencing—and participating in—noise culture.

This sensitivity, if I may say so, is required today because we tend to live numb, embedded, as we are, in our spectral age of easy image/sound production and consumption, both gluttonous and frictionless.[388] Noise culture offers a stimulus for—and way of—thinking and feeling against easy answers that never interrogates, for noise art emphasizes disorientation for the inner life. Thus, the art of noise may act as punctum in the slick palimpsest simulation in which we live, disrupting all plodding, dehumanized, routined conceits.

Dare I say it: the art of noise extends the possibility of a transforming rupture (something renewed and renewing) by addressing the frissure between intellect and the sensible. The frissure that noise art offers culture is that of a different view of the sensible, one that no longer regards the sensible as only an image (signal) cast by a remote and detached intelligibility. In noise art sophistication, signal (foreground) and noise (ground) are impenetrably interlocked and inter-embedded. And this interpenetration reveals the truth of reversibility in our culture, laced, as it is, with the counter-force of incoherence. Yet its inclination surpasses simple nihilism (as demonstrated in the Apse of Lascaux) by a collected inwardness that says a delicate *yes* to incoherent sense and impulse—and so inverts aesthetics, bending it towards rapturous plenitude.

This reciprocity of rupture and rapture is of utmost consequence. The rupture of noise is the embedded, immersive and immanent space from which the signal comes and to where it goes. Noise in art deconstructs signalness/thingness by functioning as a self-withholding ground for signal: something raw beneath and beyond conceptual language.[389]

But noise has no inherent value. It can be awful for you, or grand. It can be grand when it reminds us of the marvellous: that pre-eminent primal energy that surrounds and forms us, both beneath and beyond us—and when it de-metaphors our techno-mechanical society.[390] But mostly, it is gradational and, as such, a conceptual tool for the judicious revolutionary: those that coordinate reason and irrationality, harmony and dissidence, lucidity and obtuseness in the interests of open-minded transformation.

sO
tOday

fully emplOy randOm pOsitiOns
chOOse the kingdOm Of nOise afflictiOns
emplOy bOdily recOnfiguratiOns

embrace randOm afflictiOn frictiOns
endure all frictiOn OppOsitiOns
masquerade yOurselves in visiOns

tOy with cOnstant afflictiOn respOnses
prOccess randOm enOrmOus assumptiOns
emplOy tandOm cOntact cOnfiguratiOns
endure nOisy quest attentiOns

cOnstantly

Notes

Introduction

1. Torben Sangild points out in his essay "The Aesthetics of Noise" that, etymologically, the term "noise" in different Western languages (støj, bruit, Geräusch, larm etc.) refers to states of aggression, alarm and tension, and to powerful sound phenomena in nature such as storm, thunder and the roaring sea. It is worth noting in particular that the word "noise" comes from Greek *nausea*, referring not only to the roaring sea, but also to seasickness, and that the German *Geräusch* is derived from rauschen (the sough of the wind), related to Rausch (ecstasy, intoxication). http://www.ubu.com/papers/noise.html [accessed 1/15/2008]
2. See Luigi Russolo's seminal text *The Art of Noises* (1913) (Hillsdale: Pendragon, 2005).
3. Ideally, communication must be separated from noise. Noise is what is not communicated; it is just there as a kind of chaos, as the empirical third element of the message, the accidental part, the part of difference that is excluded.
4. Here I will focus on cultural virtues that cut against the grain and provide the grain.
5. Gilles Deleuze, "Postscript on Control Societies," in *Negotiations: 1972-1990* (New York: Columbia, 1995) 178.
6. Signal-to-noise ratio (often abbreviated SNR or S/N) is an electrical engineering measurement, also used in other fields (such as scientific measurements, biological cell signaling), is defined as the ratio of a signal power to the noise power corrupting the signal. In less technical terms, signal-to-noise ratio compares the level of a desired signal (such as music) to the level of background noise. The higher the ratio, the less obtrusive the background noise is. In image processing, the SNR of an image is usually defined as the ratio of the mean pixel value to the standard deviation of the pixel values.
7. In an interview Deleuze explains that the "key thing may be to create vacuoles of noncommunication, circuit breakers, so we can elude control". Gilles Deleuze, "Control and Becoming," trans. Martin Joughin, *Negotiations: 1972-1990* (New York: Columbia University Press, 1995) 175.
8. Normal noise, as opposed to art noise, doesn't mean anything and isn't about anything; it just is annoyingly so.
9. Guy Debord, *The Society of the Spectacle* (Detroit: Black and Red, 1976).
10. After all, each of us must make decisions about screening the wanted from the unwanted and distinguishing the essential from the random.

11. David Chalmers, "Facing up to the Problems of Consciousness," *Journal of Consciousness Studies: Controversies in Science and the Humanities* 2.3 (1995): 200-19; 208-10.

12. Noise art presents us with possible saturating border experiences in that noise is a modality of modern communication systems that is by definition *non-signifying* and deals with *signals*, and not *signs*.

13. Maria L. Assad, *Reading with Michel Serres* (Albany: SUNY Press, 1999) 18.

14. Available at http://www.archive.org/details/ViralSymphony

15. Hear in electronica music the unruly fuzziness of the *fragile* gestures of Pole, Boards of Canada, Oval, Christian Fennesz and/or Microstoria (Markus Popp of Oval and Jan Werner of Mouse on Mars) for example.

16. Impure and irregular sounds that are not tones.

17. The crux of the matter is in differentiating the difference between stimulation and wonder.

18. Chalmers, 200.

19. Gilles Deleuze, *Spinoza: Practical Philosophy* (San Francisco: City Lights, 1984) 21.

20. Thomas Metzinger, ed. *Conscious Experience* (Paderborn: Schöningh, 1995) 30.

21. As I mention phenomenology here in passing, I shall briefly relate what it is. Fundamentally, phenomenology is a philosophy of experience but the term phenomenology is often used in a general sense to refer to "subjective" experiences of various types; thus it becomes relevant to an investigation of immersive artistic states, insofar as it is a descriptive science that covers the chief features of experience taken as a whole. It is, in this sense, the study of all possible appearances in human experience during which considerations of so-called objective reality and of purely subjective response are temporarily left out of account. In the philosophical sense, phenomenology begins to redress the alienation between objectivity and subjectivity as initiated by Kant in his *Critique of Pure Reason* where Kant proposed that humans do not see the world objectively but rather through a number of ideal and subjective theory-laden categories. Philosophical systems and aesthetic theories receive their standing as truthful and useful abstractions through the human experience of the phenomenological relationship to the world. More narrowly, phenomenology is a school of philosophy whose principal purpose is to study the phenomena of human experience while attempting to suspend all consideration of objective reality or subjective association. Historically, phenomenology is the philosophical movement initiated by the German philosopher (and teacher to Aaron Gurvitch) Edmund Husserl (1859-1938) in circa 1905 and his systematic study of consciousness from a first-person standpoint. Husserl's crucial contribution to philosophy was his methodical disclosure of how meaning emerges in our consciousness of the world by our becoming conscious of our internal rapport with the world. What is relevant to this discussion is Husserl's formation of a new field of experience, the field of transcendental subjectivity which, according to Husserl, incorporates a method of access to the transcendental-phenomenological sphere in which Husserl claimed his transcendental idealism advanced beyond common idealism, beyond common realism, and beyond the very distinction between these two ideas. With the advent of phenomenology, rigorous studies of the working of consciousness were undertaken, most noticeably by the French philosopher Henri Bergson (1859-1941).

22. Primarily tone-research that led to the introduction of noise as a musical possibility.

23. In his article, "Noise as Permanent Revolution", Ben Watson points out that Ludwig van Beethoven's *Grosse Fuge* (1825) "sounded like noise" to his audience at the time. Beethoven's publishers persuaded him to remove it from its original setting as the last movement of a string quartet. He did so, replacing it with a sparkling *Allegro*, and they subsequently published it separately. See "Noise as Permanent Revolution: or, Why Culture is a Sow Which Devours its Own Farrow" in Anthony Iles and Mattin Iles, eds. *Noise & Capitalism* (Donostia-San Sebastián: Arteleku Audiolab (Kritika series), 2009) 109-10.

24. See the list of noise musicians I have been helping to maintain at http://en.wikipedia.org/wiki/List_of_noise_musicians

25. Where all can be noise, as well as noise itself being the message.

26. I do not agree with Ray Brassier in his essay, "Genre is Obsolete," when he claims that "'Noise' has become the expedient moniker for a motley array of sonic practice—academic, artistic, countercultural—with little in common besides their perceived recalcitrance with respect to the conventions governing classical and popular musics". *Multitudes*, 28 (Spring 2007).

27. Gilles Deleuze, *Spinoza: Practical Philosophy* (San Francisco: City Lights, 1984) 125.

28. Unsystematic activity at the molecular level suggests that the universe consists primarily of processes of noise.

29. Nature's difference from human representational languages.

30. Aspects of uncertainty via excess equals increased information in this model—even as I recognize that the validity of a total anything came under severe attack with post-modernism/post-structuralism in which the realization emerged that concepts and images were always already laden with specific cultural values and implicated in networks of prejudiced and invested power. Structuralist concepts of totality (based primarily on the work of the Swiss linguist Ferdinand de Saussure (1857-1913), led post-structuralists to formulate such theories, where the impossibility of ever adopting one transcendent meaning is maintained and this trend has carried over into a general inclination.

31. In *The Allure of Machinic Life*, John Johnston points to this required totalizing by recounting that while noise is often considered as coming externally into a non-noise entity (that ideally performs a clean communicative function), noise can also be viewed diagrammatized as an integral part of the same function of that system and so accorded a position *within* the diagrammatic structure (instead of residing as unmixed noise outside the communication performance). John Jonhston, *The Allure of Machinic Life. Cybernetics, Artificial Life, and the New AI* (Cambridge: MIT Press, 2008) 136-37.

32. Fundamental psychology breaks consciousness into two essential categories: the state of awareness and the subjective aspect of neurological activity (i.e., the impression of self so produced, whatever its actual cause). There are sub-categories and variations of these however. For example, some researchers define consciousness as the totality of experience at any given instant, as opposed to mind, which is the sum of all past moments of consciousness. Friedrich Wilhelm Josef von Schelling (1775-1854), in agreement with Immanuel Kant (1724-1804), maintained that the only thing we can have direct knowledge of is our consciousness. However, consciousness, in Aldous Huxley's (1894-1963) view (as influenced by William James's (1842-1910)

study *The Varieties of Religious Experience: A Study in Human Nature*), is mainly an abridgement application that allows us to construct a coherent world view based on selective oblivion. Aldous Huxley (1970) 22. Brian Massumi upheld Huxley's/James's "subtractive" understanding of consciousness by seeing both will and consciousness as "limitative, derived functions which reduce a complexity too rich to be functionally expressed". Brian Massumi, "The Autonomy of Affect," *Cultural Critique* (Fall 1995) 83-109; 90.

33. Librarian, libertine, paleologist, archivist, radical thinker, author of erotic fiction; Bataille took an active role in the mid-20th century Parisian avant-garde art and literary scene by objecting to what he saw as the aestheticism and sentimentality of the Surrealists. Consequently he became André Breton's (1896-1966) antagonist from the intellectual ultra-left. After World War II, as founding editor of the journal *Critique* and after authoring the transgressively philosophical books *L'Expérience Intérieure* (Inner Experience) (1943), *Le Coupable* (Guilty) (1944), *Sur Nietzsche* (On Nietzsche) (1945) and *La Part Maudite* (Accursed Share) (1947), Bataille's thought emerged as a viable alternate to Jean-Paul Sartre's then reigning philosophical school of Parisian Existentialism.

34. For Georges Bataille, examples of non-productive excess/expenditure can be found (in varying degrees) in forms of luxury, lamentation, spectacle, art, poetry, erotic activity and mystical endeavours; some of which place an emphasis on a loss that must be as great as possible in order for that activity to take on its fullest meaning. For the finest comprehensive overview of Bataille's thought in this regard, see his book *Eroticism*, trans. Mary Dalwood, intro. Colin MacCabe (London: Penguin, 2001), along with Denis Hollier's book on Bataille's general postulates, *Against Architecture* (Cambridge: MIT Press, 1990).

35. As is art. Piet Mondrian has made the valid point that fine art is not made for anybody and is, at the same time, for everybody.

36. I should establish that the pantheoristic definition of noise in art which I am upholding here, and which I find requires reiteration as artists move increasingly from organic materials to the use of electronic and synthetic ones, is basically that supplied by Susanne Langer in her book *Feeling and Form* where she determines that "art is the creation of forms symbolic of human feeling," Susanne K. Langer, *Feeling and Form* (NJ: Prentice Hall, 1953) 40.

37. Indeed Leon Cohen in his paper "The History of Noise" makes a case for noise being involved in the solution of some key scientific, mathematical and technological problems.

38. When I use the terminology *expanded* here I am referring to the rich meaning given to it by Gene Youngblood in his book *Expanded Cinema* as that which transgresses and exceeds the customary boundaries of our encounters. When Youngblood discusses what he calls "expanded cinema" he refers it to an "expanded consciousness," Gene Youngblood, (New York: E. P. Dutton and Co, Inc. 1970) 41.

39. In science, and especially in physics and telecommunication, noise is fluctuations in and the addition of external factors to the stream of target information (*signal*) being received at a detector. White noise is always present.

40. Allen S. Weiss, *Phantasmic Radio* (Durham: Duke University Press, 1995) 90.

41. In philosophic terms, subjectivity denotes how the truth of some privileged class of statements depends on the mental state or reactions of the person making the

statement. In epistemology, subjectivity is knowledge that is restricted to one's own perceptions. This implies that the qualities experienced by the senses are not something belonging to the physical beings, but are subject to interpretation. In metaphysics, subjectivity includes the idea of solipsism. In aesthetics, subjectivism is the view that statements about beauty (for example) are not reports of "objective" qualities inherent in things but rather cognitive reports of internal feelings and attitudes.

42. The concept of the sublime as such first emerged as the topic of an incomplete treatise entitled "On the Sublime" that is believed to have been written in the mid-third century AD by Cassius Longinus (third century AD). The author of the treatise defines sublimity as: (1) excellence in language, (2) the expression of a great spirit, and (3.) the power to provoke ecstasy. The immersive sublime, from the combined point of view of Cassius Longinus's last two definitions, is apprehended and grasped as a totality while at the same time experienced as exceeding our usual lucidity, therefore provoking a sensation of awe (as recognized by Longinus). Centuries later, the term was given special prominence by Edmund Burke (1729-1797) in his *A Philosophical Enquiry Into the Origin of Our Ideas of the Sublime and Beautiful* (1757), one of the most popular 18th century treatises on aesthetics, which was translated into French in 1765 and into German in 1773. According to Burke, the sublime feeling (which contrasts with that of the beautiful) is caused by a mixture of terror, admiration, apprehension, and supra-attention. Burke maintained that the life of the spirit depends on this type of awe in agreement with the immense scheme of the universe.

43. Romanticism (circa 1795-1840) is the cultural movement inspired by the writings of Edmund Burke (1729-1797) and the French philosopher Jean Jacques Rousseau (1712-1778), (among others) that focused on individual passions and inner struggles and hence produced a new outlook and positive emphasis on the emotional artistic imagination which became perceived as a gateway to transcendent experiences of unity.

44. Torben Sangild points out in his essay "The Aesthetics of Noise" that in *Genèse*, French philosopher Michel Serres sketched out the idea that the ultimate being-in-itself is noise. Behind the phenomenal world (the world we perceive)—he proposes—is an infinite complexity, an incomprehensible multitude analogous to white noise. (http://www.ubu.com/papers/noise.html accessed 1/15/2008). What Serres initially finds intriguing about noise (rather than the message) is that it opens up a fertile avenue of reflection. Instead of remaining pure noise, it becomes a means of transport. Also, Serres addresses the theme of noise and communication to show that 'noise is part of communication'; it cannot be eliminated from the system.

45. Jane. Ellen Harrison, *Ancient Art and Ritual* (Bradford-on-Avon, Wilts: Moonraker Press. 1913) 80.

46. In information theory, entropy is a measure of the uncertainty associated with a random variable.

47. Harsh feedback sounds are tones that may have drone-like charactistcs or swarm chaotically.

48. Bataille has remained a great fascination, though always slightly out of distance, given his penchant for violence, as I am largely a pacifist, politically. I believe in the political effectiveness of civil disobedience and passive *résistance*. So he is always pithily slipping from my mental fingers. But with his last book, *The Tears of Eros*, he provided the inspiration for my art to attempt a visual sagacity that tests the limits of form and

stretches the bounds of meaning by recasting our experiences of encountering wildly disjunctive phantasmagoric data on the Internet into the sumptuous physicality of negation. In that sense, he turned my work against the grain of its prior obsession with fabricating a complicated forensic fairy-tale out of the internet's grisly *mélange*, a *mélange* which keeps slipping in and out of idiosyncratic narration as it keeps folding and unfolding. When I went to Vézelay to visit his tomb, I searched the cemetery for something like two hours, reading each headstone meticulously. But I was unable to discover his grave! *Merde!* I had hoped to leave a little perverse poem and perhaps defecate on it, but no luck. In that sense I am reminded that frustration often amplifies desire and that this is essential to noise in art also.

49. Georges Bataille, *Visions of Excess* (Minneapolis: University of Minnesota Press, 1985) 116-29.

50. Otto Kernberg, *Borderline Conditions and Pathological Narcissism* (New York: Jason Aronson. 1975) 26.

51. Noise vs. music, non-intended sounds vs. intended sounds, life vs. art; the oppositional pairs resonating along with the first opposition form an ever-extending thread.

52. Stephen Talbott, *The Future Does Not Compute: Transcending the Machines in Our Midst* (Sebastopol: O'Reilly and Associates, 1995).

53. Gene Youngblood, *Expanded Cinema* (New York: E. P. Dutton and Co, Inc. 1970) 81.

54. Youngblood, 85.

55. Anton Ehrenzweig, *The Hidden Order of Art* (Berkeley: University of California Press, 1967) 9.

56. Thomas Metzinger, ed. *Conscious Experience* (Paderborn: Schöningh, 1995) 14.

57. For a consideration of aural history see Emily Thompson, *The Soundscape of Modernity Architectural Acoustics and the Culture of Listening in America, 1900-1933* (Cambridge: MIT Press, 2004).

58. Metzinger, 15.

59. This concept owes something to Quentin Meillassoux's idea of *hyper-Chaos* that was sketched out in *After Finitude* (64): a form of absolutization where nothing is impossible or unthinkable. Quentin Meillassoux, *After Finitude: An Essay on the Necessity of Contingency*, trans. Ray Brassier (London: Continuum, 2008).

60. Given our heightening condition of connectivity, the heterogeneous, multiplicitous, spreading and non-hierarchical nature of the epistemological rhizome come together under the hyper (i.e. connected) effect of hyper-noise.

61. We know from Michel Foucault (1926-1984) how all ideals, all symbols in fact, can be readily adapted to fit the dictates of social power. Surely incoherent views of the whole have been destructive, as Boris Groys's book *The Total Art of Stalinism: Avant-garde, Aesthetic Dictatorship and Beyond* (Princeton: Princeton University Press, 1992) substantially makes apparent. However, we also know that ideals are indispensable in creating possibilities for change.

62. John Cage, *Silence* (London: Calder and Boyers, 1966) 14.

63. Excess noise radiation indicates that the universe is continuously expanding.

64. Wikipedia defines as Noology or Noölogy deriving "from the Greek words νοῦς 'mind' and λόγος 'logos'. Noology thus outlines a systematic study and organization of everything dealing with knowing and knowledge, i.e.: cognitive neuroscience. It is also used to describe the science of intellectual phenomena. It is the study of images of thought, their emergence, their genealogy, and their creation." Available at https://secure.wikimedia.org/wikipedia/en/wiki/Noology.

65. The essential opposing philosophy to epistemological symbolization is that of rational empiricism, but there are many gradations and intermediary positions staked out from this binary opposition, including Immanuel Kant's synthetic a priori, which allowed for an account of art, among other things. Kant held that space was in essence mental and a priori to the perception of exterior objects.

66. Although Freud was not an intentional aesthetic theorist, he did influence aesthetic theory through his comprehensive psychoanalytic framework and through his use of art to elucidate the fact that the breadth of psychoanalysis extended beyond dreams and neurosis into aesthetic achievements.

67. In that regard, it is interesting that Russolo, in his manifesto of March 11, 1911 writes: "It will not be through a series of lifelike noises but through the fantastic associations of these various tones and rhythms that the new orchestra will obtain the most complex and new sound emotions".

68. Extemporaneous conceptual/cultural intoxication involving powerful auto improvisations.

69. Take, for example, *NoiseFold*, an interactive visual-music-noise performance that draws equally from mathematics, science and the visual and sonic arts. This networked performance duet explores the use of infrared and electromagnetic sensors to manipulate and fold virtual 3-D objects that emit their own sounds. The work integrates multiple techniques including real-time 3-D animation, mathematic visualization, recombinant non-linear database, A-life simulation, image to sound transcoding, complex data feedback structures and a variety of algorithmic processes used to generate both sonic and visual skins. The result is a theater of emergence and alchemical transformation existing within an intricate cybernetic system.

70. Given the impossibility of self-identical signal transmission.

Chapter 1

71. There is no opportunity to get rid of the deferring effects of noise, as it is a fundamental principle of the physical world.

72. See Alex Ross, *The Rest Is Noise: Listening to the Twentieth Century* (New York: Farrar, Straus and Giroux, 2007) on how twentieth-century composers felt compelled to create a bewildering variety of sounds that approached the purest beauty to the purest noise (even as I do not accept that dichotomy).

73. In an interview John Cage refers to a workshop he conducted: "I had the lights turned out and the windows open. I advised everybody to put on their overcoats and listen for half an hour to the sounds that came in through the window, and then to add to them—in the spirit of the sounds that are already there, rather than in their individual spirits. That is actually how I compose. I try to act in accord with the absence of my music". (Gena and Brent, 176).

74. Excess here, I wish to point out, may also be of the silent—almost unperceivable—type as well. For more on this approach see the essay "Silence is Sexy: The Other "Extreme" Music" by Thomas Bey William Bailey in his book *Micro Bionic: Radical Electronic Music And Sound Art In The 21st Century*.

75. Dadaism tries to expose the impotence of reason and technology while being aware of its social power. In an effort to achieve this goal, an illogical sound-poetry (adapted by Dadaism from Italian Futurism) was common at Zurich's *Cabaret Voltaire* (founded in 1916 by Hugo Ball Tristan Tzara, Hans Arp, Marcel Janco and Richard Huelsenbeck). Simultaneous poems by Henri-Martin Barzun and Alfred Jarry's *Ubu Roi* were recited there. For more on Dada as an ongoing active urge, see Andrei Codrescu's *Posthuman Dada Guide* (Princeton: Princeton University Press, 2009).

76. That in philosophy which is concerned with theories of knowledge.

77. Gilles Deleuze and Félix Guattari, *What is Philosophy?* (London: Verso, 1994) 7.

78. Gilles Deleuze and Félix Guattari, *A Thousand Plateaus: Capitalism and Schizophrenia* (Minneapolis: University of Minnesota Press, 1987) 21.

79. Anti-art is the definition of a work of art that may be exhibited or delivered in a conventional context but makes fun of serious art or challenges the nature of art. A work such as Marcel Duchamp's *Fountain* of 1917 is a prime example of anti-art.

80. John Cage's 'anti-art' music still operates within an aesthetic abstraction similar to art music, but the aesthetic isolation and abstraction are questioned. Thus, the borders become permeable.

81. Paul Crowther, *Critical Aesthetics and Postmodernism* (Oxford: Clarendon Press, 1993) 60.

82. John Johnston, *The Allure of Machinic Life: Cybernetics, Artificial Life, and the New AI* (Cambridge: MIT Press, 2008) 26-8.

83. Norbert Wiener, *Cybernetics or Control and Communication in the Animal and the Machine* (Cambridge: MIT Press, 1961) 42.

84. Paul Hegarty, *Noise Music: A History* (London: Continuum, 2007).

85. Douglas Kahn, *Noise, Water, Meat: A History of Sound in the Arts* (Cambridge: MIT Press, 2001).

86. Jacques Attali, *Noise: The Political Economy of Music* (Minneapolis: University of Minnesota Press, 1985).

87. Allen S. Weiss, *Phantasmic Radio* (Durham: Duke University Press, 1995) 90.

88. See *Artística de Valencia, After The Net*, 5 – 29 June 2008, Valencia, Spain catalogue: *Observatori 2008 After The Future* (80).

89. Often using scratched, warped, defective, damaged aspects of recording technology.

90. Lazlo Moholy-Nagy recognized in 1923 the unprecedented efforts of the Italian Futurists to broaden our perception of sound using noise. In an article in *Der Storm* #7, he outlined the fundamentals of his own experimentation: "I have suggested to change the gramophone from a reproductive instrument to a productive one, so that on a record without prior acoustic information, the acoustic information, the acoustic phenomenon itself originates by engraving the necessary *Ritchriftreihen* (etched

grooves)." He presents detailed descriptions for manipulating discs, creating "real sound forms" to train people to be "true music receivers and creators". See UbuWeb Papers, *A Brief history of Anti-Records and Conceptual Records* by Ron Rice.

91. "Dada applies itself to everything, and yet it is nothing, it is the point where the yes and the no and all the opposites meet, not solemnly in the castles of human philosophies, but very simply at street corners, like dogs and grasshoppers". From Tristan Tzara's "Dada Manifesto" [1918] and "Lecture on Dada" [1922], translated from the French by Robert Motherwell in *Dada Painters and Poets*, Robert Motherwell (Cambridge: Harvard University Press, 1989) 78-9. Leading Dadaists include Hans (Jean) Arp, Sophie Taeuber-Arp, Hannah Höch, Raoul Hausmann, George Grosz, John Heartfield, Kurt Schwitters, Max Ernst, Francis Picabia, Man Ray, Marcel Duchamp, Louis Aragon, Johannes Baader, Hugo Ball, André Breton, Jean Crotti, Paul Eluard, I.K. Bonset, Marcel Janco, Clément Pansaers, Tristan Tzara, Hans Richter and the lesser known – but one of my personal favourites—Baroness Elsa von Freytag-Loringhoven (née Else Plötz).

92. Mathew Biro, *The Dada Cyborg: Visions of the New Human in Weimar Berlin* (Minneapolis: University of Minnesota Press, 2009) 50. Also, I have been informed by Timothy Shipe, the curator of The International Dada Archive at The University of Iowa, that the performance of *Antisymphonie* was held at the Graphisches Kabinett, Kurfürstendamm 232, at 7:45 PM. The printed program lists 5 numbers: *Proclamation dada 1919* by Huelsenbeck, *Simultan-Gedicht* performed by 7 people, *Bruitistisches Gedicht* performed by Huelsenbeck (these latter 2 pieces grouped together under the category *DADA-machine*), *Seelenautomobil* by Hausmann, and finally, Golyscheff's *Antisymphonie* in 3 movements, subtitled *Musikalische Kriegsguillotine*. The 3 movements of Golyscheff's piece are titled *provokatorische Spritze, chaotische Mundhöhle oder das submarine Flugzeug*, and *zusammenklappbares Hyper-fis-chendur*.

93. Watts made a series of spray-painted records for a Fluxus performance at the Fluxstore on Canal Street played by the audience, and as the paint wore off, gradually the music was revealed.

94. Generally his noise music is created from damaged LP recordings: often cut and glued together or painted over or melted. Hungarian constructivist László Moholy-Nagy did similar noise experiments in the 1920s.

95. Thomas J. Harrison, *The Emancipation of Dissonance* (Berkeley: University of California Press, 1910).

96. In *Futurism and Musical Notes*, Daniele Lombardi discusses the mysterious case of the French composer Carol-Bérard; a pupil of Isaac Albeniz. Carol-Bérard is said to have composed a *Symphony of Mechanical Forces* in 1910 but little evidence has emerged thus far to establish this assertion.

97. In 1977, the Historic Archives of Contemporary Arts of the Venice Biennale organized an exhibition of Russolo's work and the curator, Gian Franco Maffina, had five noise-intoners reconstructed for the occasion. (Russolo's original instruments had all been destroyed during the Second World War). See *Futurism and Musical Notes* by Daniele Lombardi on UbuWeb.com, originally printed in *Artforum* as translated by Meg Shore.

98. In 1912, Marcel Duchamp, along with Appollinaire and Picabia, attended a performance of *Impressions of Africa*, a play by an obscure author named Raymond Roussel. Roussel greatly admired the works of the author Jules Verne, which he read

over and over again, fascinated with their extraodinary voyages and machines, full of bachelor scientists completely absorbed in positivist exploratory dreams taken to delirious extremes. Duchamp later credited Roussel with the inspiration for his *The Bride Stripped Bare by Her Bachelors, Even*. In 1912, Duchamp started producing paintings and drawings depicting mechanized sex acts such as *Mechanics of Modesty* and *The Passage from the Virgin to the Bride*. At the same point in time, Freud was explaining in his lectures that complex machines in dreams always signified the genital organs. Roussel invented language machines that produced texts through the use of repetitions and combination/permutations. This machine-like logic provided his art with a seemingly pure spectacle of an endless variety of textual games and combinations flowing in circular form. Within this writing process, Roussel described a number of fantastic machines, including a painting machine in his novel *Impressions of Africa*. This painting machine wonderfully describes and foresees the arrival of computer-robotic technology and its application to visual art which we have available to us today, nearly a century after he envisioned it. Thus it is through Roussel that we might start to map a certain lineage in the avant-garde noise through out our century, passing through Duchamp and the Futurists.

99. For a fascinating discussion of Cage's *Imaginary Landscape* works in relationship to noise, see Allen S. Weiss, *Phantasmic Radio* (Durham: Duke University Press, 1995) 49-55 with a focus on *Imaginary Landscape #4*, 54-55.

100. Weiss, 45-52.

101. Weiss, 9-34.

102. The original Dada model for these art startegies are beautifully exemplified by the early photomontages of Hannah Höch (see Matthew Biro, *The Dada Cyborg: Visions of the New Human in Weimar Berlin* (Minneapolis: University of Minnesota Press, 2009) 65-103, and the assemblage *God* (c. 1917) by Baroness Elsa von Freytag-Loringhoven and Morton Livingston Schamberg that is in The Philadelphia Museum of Art's Louise and Walter Arensberg Collection. Dada questioned and affected what art can look like, as well as what art can do, and set the stage for many avant-garde movements, including Surrealism, Pop Art, Performance Art and Digital Art. Dada also irrevocably changed the landscape of popular culture, influencing graphic design, advertizing, and film.

103. *Barbarella* is a 1968 erotic sci-fi film staring Jane Fonda directed by Roger Vadim based on the French *Barbarella* comics of Jean-Claude Forest. By appropriately manipulating the keys of the *Excessive Machine*, a player of this torturous musical instument may induce enormous sexual pleasure, sufficient to cause death by orgasm. In one of the final scenes of the movie, the evil opponent is torturing Barbarella with the pleasures of this machine, but in the end the machine overloads and is destroyed in a burst of noise. Barbarella survives and feels rather grand.

104. See Paul Hegarty's essay *Full With Noise: Theory and Japanese Noise Music* http://www.ctheory.net/articles.aspx?id=314 [accessed 28 October, 2010].

105. Caleb Kelly, *Cracked Media: The Sound of Malfunction* (Cambridge: MIT Press, 2009) 6-24.

106. Kim Cascone, "The Aesthetics of Failure: 'Post-Digital' Tendencies in Contemporary Computer Music," *Computer Music Journal* 24.4 (Winter 2002): 12–18.

107. Steve Goodman, *Contagious Noise: From Digital Glitches to Audio Viruses* in Jussi Parikka, and Tony D. Sampson (eds.) *The Spam Book: On Viruses, Porn and*

Other Anomalies From the Dark Side of Digital Culture (Cresskill, NJ: Hampton Press, 2009) 128.

108. In that noise may instigate a decentering of subjectivity.

109. *Mona Lisa Overdrive* is a cyberpunk novel by William Gibson published in 1988 and the final novel of the *Sprawl* trilogy, following *Neuromancer* and *Count Zero*.

110. Donald Lowe, *History of Bourgeois Perception* (Chicago: University of Chicago Press, 1982) 18-23; 26.

111. Op Art: a hard-edge geometrical movement which flourished in the early-1960s largely inspired by various optical experiments of Marcel Duchamp.

112. According to Arthur Koestler's holon concept (established in *Beyond Reductionism* and in *The Ghost in the Machine*, 1967, 45-58), instead of cutting up immersive perceptual wholes into discrete focal parts, immersion should be scrutinized and understood using synthetic sub-whole sets found within the ambient atmospheric spectrum of immersive perception's entirety. It is the exposé of the synthetic atmospheric phenomenology of such holonertic listening (dependent on the linked and amassed sum-total) that will concern us here.

113. Arthur Koestler, *The Ghost in the Machine* (New York: Hutchinson, London and Macmillan, 1967) and *Beyond Reductionism* (Boston: Beacon, 1971).

114. Koestler, *The Ghost in the Machine*, 45-58.

115. *XS: The Opera Opus* was a no wave avant-garde music and art performance created by Rhys Chatham and Joseph Nechvatal in the mid 1980s. Jane Lawrence Smith sang the lead role in the Boston performance and Yves Musard danced the main role. Its theme was the excess of the nuclear weapon buildup of the Ronald Reagan presidency. *XS: The Opera*, Shakespeare Theater, Boston was the final production and consisted of three soprano singers, 4 trumpets, six electric guitars, base, drums, 35 mm slide projection and dance. The duration was 90 minutes. Choreography: Yves Musard, 35 mm cross-fade art slides: Joseph Nechvatal. See Rosetta Brooks, "Interview-Rhys Chatham & Joseph Nechvatal", *ZG*, (#12 Fall) and/or Robert Kleyn, "The Shadow Reflected", *ZG* (#12 Fall).

116. *Tellus* was created in 1983 in New York City by myself, Claudia Gould, a curator, and Carol Parkinson, a composer and staff member of Harvestworks/Studio PASS. We began to collect, produce and document the art of audio through publishing works from local, national and international artists. We worked with contributing editors who proposed themes and collected the best works from that genre. Unknown artists were teamed with well-known artists; historical works were juxtaposed with contemporary and high art with popular art, all in an effort to enhance the crossover communication between the different mediums of art—visual, music, performance and spoken word. Tellus is archived on the web at http://www.ubu.com/sound/tellus.html

117. http://www.ubu.com/sound/tellus_13.html

118. http://www.ubu.com/sound/tellus_20.html

119. See, for example, Chicago Museum of Contemporary Art, *The Record as Artwork from Futurism to Conceptual Art. The Collection of Germano Celant*. Exhibition Catalog, 1977.

Chapter 2

120. Deleuze, and Guattari, *On The Line* (New York: Semiotext(e), 1983) 2.

121. Gene Youngblood, *Expanded Cinema.* (New York: E. P. Dutton and Co, Inc., 1970) 41.

122. Non-communication.

123. That is, disorder, chance, and the exceptional.

124. Theodor Adorno, *Aesthetic Theory* (London: Routledge and Kegan Paul, 1984) 271.

125. R. G. Collingwood, *Principles of Art* (Oxford: Claredon Press, 1938) 223.

126. Robert Stewart, ed. *Ideas That Shaped Our World: Understanding the Great Concepts of Then and Now* (London: Marshall, 1997) 93.

127. Janet Wolff, *The Social Production of Art* (NY: New York University Press, 1993) 1.

128. Wolff, 92.

129. Teresa Brennan, and Martin Jay, eds. *Vision in Context* (London: Routledge, 1996) 31.

130. Hal Foster, ed. *Vision and Visuality* , (Seattle: Bay Press,1988) x.

131. Guy Debord, *The Society of the Spectacle* (Detroit: Black and Red, 1976).

132. Jacqueline Rose, *Sexuality in the Field of Vision* (London: Verso, 1986).

133. Michel Foucault, *Discipline and Punish: The Birth of the Prison* (New York: Vintage, 1979) 201. According to Foucault, the major effect of the panopticon (a circular prison designed by the British philosopher Jeremy Bentham (1748-1832) based on his principles of "happiness calculus") is to induce in the prison inmate (and by extension anyone) a state of consciousness that assures the automatic functioning of power.

134. Jane Ellen Harrison, *Ancient Art and Ritual* (Bradford-on-Avon, Wilts: Moonraker Press, 1913) 21.

135. Homo Sapiens with large frontal-lobes who migrated from the Middle East about 17, 000 BP. At first their art consisted of intimate body decoration such as beads, bracelets, pendants and necklaces.

136. L'Oreille d'Enfer in Les Eyzies-de-Tayac-Sireuil and *Grotte de Pair-non-Pair* in Gironde are good examples.

137. Brigitte Delluc, and Gilles Delluc, *Discovering Lascaux* (Pollina à Luçon: Editions Sud Ouest, 1990) 46.

138. Delluc and Delluc, 55.

139. Carla Hoekendijk, ed. *Interfacing Realities* (Amsterdam: V2 Organisatie, 1997) 21.

140. Howard Rheingold, *Virtual Reality* (New York: Summit Books, 1991) 379-82.

141. Alfred Kinsey, et al., *Sexual Behavior in the Human Female* (Philadelphia: Saunders, 1953) 615.

142. Begining about 45,000 to 38,000 years ago and ending around 10,000.

143. Delluc and Delluc, 57.

144. If one accepts the point concerning an infant's rapport with voluminous breasts.

145. The concept of the *Gesamtkunstwerk* (total-artwork) is a proposition rooted in the Neo-Platonic heritage of Romanticism. However, for Richard Wagner, it took on a narrow and precise meaning as he re-theorized it in his 1849 hypothetical essays "The Artwork of the Future" (Das Kunstwerk der Zukunft) and "Art and Revolution" (Kunst und Revolution). "The Artwork of the Future" has two principal themes. The first proclaims the doctrine of an "art of the people" which idealized art in a way that would necessarily engage the masses (inasmuch as it was a narration of the masses own thoughts, feelings and aspirations) as Wagner had imagined existed during the period of the Greek dramas. This is the *Gesamtkunstwerk* ideal in the kind-hearted political sense. In order to attain this level of idealized democratic-communist amiable social blend, the formal characteristics of the *Gesamtkunstwerk* were theorized as necessarily being the product of a fusion of the separate arts in pursuit of a "total effect" which would be achieved through a total synthesis in which all of the individual arts contribute.

146. Officially discovered by Emile Cartailhac in 1906.

147. Mario Ruspoli, *The Cave of Lascaux: The Final Photographic Record* (New York: Abrams, 1987) 82.

148. Georges Bataille, *Oeuvres Completes: Lascaux: La Naissance de l'Art* (Paris: Gallimard, 1979) 46.

149. Carbon dated circa 17, 000 BP.

150. Georges Bataille, *Oeuvres Completes: Lascaux: La Naissance de l'Art* (Paris: Gallimard, 1979) 37.

151. Some of the paintings in Chauvet cave are over 30, 000 years old, 3, 000 years older than the oldest cave paintings previously known and nearly twice as old as those found at Lascaux.

152. Mario Ruspoli, *The Cave of Lascaux: The Final Photographic Record* (New York: Abrams, 1987) 81.

153. Full of interruption and corruption and ruptures of information.

154. Brigitte Delluc, and Gilles Delluc, *Discovering Lascaux* (Pollina à Luçon: Editions Sud Ouest, 1990) 74.

155. Steven Mithen, *The Prehistory of the Mind: The Cognitive Origins of Art, Religion and Science* (Cambridge: Cambridge University Press, 1996) 165-67.

156. John C. Lilly, *Programming and Metaprogramming in the Human Bio-Computer: Theory and Experiments* (New York: Bantam Books, 1974) 40.

157. Deleuze, and Guattari, *Nomadology: The War Machine* (New York: Semiotext(e), 1986) 13.

158. Deleuze, and Guattari, *Nomadology* , 36.

159. Deleuze and Guattari, *Nomadology,* 36.

160. André Leroi-Gourhan, *The Art of Prehistoric Man in Western Europe* (London: Thames and Hudson, 1968) 163.

161. Mario Ruspoli, *The Cave of Lascaux: The Final Photographic Record* (New York: Abrams, 1987) 80.

162. Leroi-Gourhan, 315.

163. Mario Ruspoli, *The Cave of Lascaux: The Final Photographic Record* (New York: Abrams, 1987) 146-47.

164. Ruspoli, 146.

165. Ruspoli, 171.

166. Georges Bataille, *Oeuvres Completes: Lascaux: La Naissance de l'Art* (Paris: Gallimard, 1979) 58-9.

167. Leroi-Gourhan, 315.

168. Albert Rothenberg, *The Emerging Goddess* (Chicago: University of Chicago, 1979) 342.

169. Bataille, 59.

170. Leroi-Gourhan, 316.

171. Brian Massumi, "The Autonomy of Affect," *Cultural Critique* (Fall 1995): 83-109 (91).

172. Hegel's notion of the absolute consisted of becoming other in spirit.

173. Richard Rudgley, *The Alchemy of Culture: Intoxicants in Society* (London: British Museum Press, 1993) 28.

174. Deleuze and Guattari, *A Thousand Plateaus: Capitalism and Schizophrenia* (Minneapolis: University of Minnesota Press, 1987) 510.

175. Deleuze and Guattari, *A Thousand Plateaus* , 153.

176. Brian Massumi, *A User's Guide to Capitalism and Schizophrenia: Deviations from Deleuze and Guattari* (Cambridge: MIT Press, 1992) 70.

177. Also further altered by the meta-nihilistic chaos of repressed excess.

178. Leroi-Gourhan, 33.

179. This idea ties into *Simultaneity* in music: attempts at interweaving sound fragments.

180. Transmission as Derridean *différance*.

181. Nigel Pennick, *The Ancient Science of Geomancy* (London: Thames and Hudson, 1979) 119.

182. Susanne Langer, *Feeling and Form: A Theory of Art Developed from Philosophy in a New Key* (New York: Charles Scribner's Sons, 1953) 69.

183. *Grotte* in French and *grotta* in Italian.

184. Naomi Miller, *Heavenly Caves: Reflections on the Garden Grotto* (London: Allen and Unwin, 1982) 17.

185. Miller, 18-20.

186. Miller, 7.

187. Allen S. Weiss, *Mirrors of Infinity: The French Formal Garden and 17th Century Metaphysics* (New York: Princeton Architectural Press, 1995a) 48.

188. Jacques Bersani et al. eds. *Archéologie: Les Grands Atlas Universalis* (Paris: Encyclopedia Universalis, 1985) 38-9.

189. De Antro Nympharum is a consequential and elaborate interpretation and defence of Paganism, which adapted Plotinus's (AD 205-270) teachings while putting extra emphasis on the importance of theurgic magical practices. In turn, Porphyry's theurgic theories were succeeded by those of Iamblichus (circa AD 250-312) (who also emphasized the preternatural theurgic factors in Neo-Platonism) and Jamblichus (circa AD 255-315) who also maintained a belief in sorcery and theurgy (the art of compelling demons and other supernatural powers to produce desired results).

190. Porphyry, *L'Antro des Nymphes* (Paris: Pléiade, 1918) 20.

191. Vecellio Titien's (1488-1576) painting from the Renaissance era, *Nymph and Shepherd* (1576), now at the Kunsthistorisches Museum in Vienna, illustrates the nymph concept beautifully.

192. Miller, 13.

193. Depicted in human form with the legs, horns and ears of a goat.

194. Miller, 15.

195. Jane E. Harrison, *Ancient Art and Ritual* (Bradford-on-Avon, Wilts: Moonraker Press, 1913) 64.

196. Emmanuel Levinas, *Totality and Infinity: An Essay on Exteriority* (Pittsburgh: Duquesne University Press, 1969) 72-7.

197. *Viractualism* is an art theory term I developed in 1999. The term viractualism (and *viractuality*) emerged out of my doctoral research into the philosophy of art and new technology concerning immersive virtual reality at Roy Ascott's Centre for Advanced Inquiry in the Interactive Arts (CAiiA), University of Wales College, Newport, UK (now the Planetary Collegium at the University of Plymouth). There I developed this concept of the viractual, which strives to create an interface between the biological and the technological. Viractualism is central to my work as an artist. The basis of the viractual conception is that virtual producing computer technology has become a noteworthy means for making and understanding contemporary art and that this brings artists to a place where one finds the emerging of the computed (the virtual) with the uncomputed corporeal (the actual). This amalgamate—which tends to contradict some central techno clichés of our time—is what I call the viractual. Digitization is a key metaphor for viractuality in the sense that it is the elementary translating procedure today. For me, the viractual recognizes and uses the power of digitization while being culturally aware of the values of monumentality and permanency—qualities that can be found in some compelling analog art. A key influence on me was Gilles Deleuze's consideration of Baruch Spinoza, the 17th century philosopher who merged mind and matter into one material. In his "Spinoza: Practical Philosophy," Deleuze pointed me towards an acknowledgment of desires' productiveness, as Deleuze indicated how desires drive us to stir towards greater or lesser states of exalted comprehensiveness depending on whether the thing encountered enters into composition with us, or on the contrary, tends to decompose us. I think that viractualism signals a new sensibility emerging in art respecting the integration of certain aspects of science, technology, myth and consciousness—a consciousness struggling to attend to the prevailing contemporary spirit of our age in which everything, everywhere, all at once is connected in a rhizomatic web of transmission. But the viractual realm is also a political-spiritual chaosmos in the sense

that new forms of order may come up such that any form of order is only temporary and provisional. Within viractual creation, all signs are subject to boundless semiosis, which is to say that they are translatable into other signs. Here, of course, it is possible to find resonances and affinities between formal and conceptual opposites. I suggest that the term (concept) viractual (and *viractualism* or *viractuality*) may be an entrainment/*égréore* conception helpful in defining our now third-fused interspatiality which is forged from the meeting of the virtual and the actual, a concept close to what the military call augmented reality, which is the use of transparent displays worn as see-through glasses on which computer data is projected and layered.

198. Harrison, 65.

199. Harrison, 66.

Chapter 3

200. Robert Romanyshyn, *Technology as Symptom and Dream* (London: Routledge, 1989) 83-93.

201. Romanyshyn, 33.

202. Romanyshyn, 33.

203. Samuel Edgerton, *The Renaissance Discovery of Linear Perspective* (New York: Harper and Row, 1976) 119.

204. William Ivins, *Art and Geometry* (New York: Dover, 1964) 69.

205. Romanyshyn, 97-101.

206. Zenon Pylyshyn, "Here and There In the Visual Field" in Zenon Pylyshyn, ed. *Computational Processes In Human Vision: An Interdisciplinary Perspective* (Norwood, NJ: Ablex, 1988) 210-38.

207. Romanyshyn, 42.

208. Croix de la Horst and Richard Tansey *Gardner's Art through the Ages* (New York: Harcourt, Brace and Javanovich. 1975) 433.

209. Romanyshyn, 77.

210. Martin Kemp, *The Science of Art: Optical Themes in Western Art from Brunelleschi to Seurat* (New Haven: Yale University Press, 1990) 167-220.

211. Erwin Panofsky, *Tomb Sculpture: Four Lectures on its Changing Aspects from Ancient Egypt to Bernini* (New York: Abrams, 1991) 34.

212. Interestingly Michel Serres delivered a speech "Physical and Social Sciences: The Case of Turner," based on two paintings by Turner at the LSU College of Art and Design.

213. Romanyshyn, 216-21.

214. A Florentine architect and engineer, and mastermind of the distinctive dome that crowns the cathedral in Florence.

215. Hal Foster, ed. *Vision and Visuality* (Seattle: Bay Press, 1988) 7.

216. Edgerton, 9.

217. Romanyshyn, 71-82.

218. Romanyshyn, 74.

219. John Berger, *Ways of Seeing* (New York: Penguin Books, 1977) 109.

220. Michael Schuyt, Joost Elffers, and George Collins, *Fantastic Architecture: Personel and Eccentric Visions* (New York: Harry Abrams, 1980) 177.

221. Joseph Nechvatal, "Hyper-death and the Scopic Corpse," *Artforum* (November, 1990): 130.

222. Joseph O'Neil, ed. *Mexico: Splendors of Thirty Centuries* (New York: Metropolitan Museum of Art, 1990) 27.

223. O'Neil, 358-59.

224. Madelene Ostwald, "Structuring Virtual Urban Space: Aboroscent Schemes" in Peter Droege, ed. *Intelligent Environments: Spatial Aspects of the Information Revolution* (Amsterdam: Elsevier, 1997) 451- 82; 457.

225. Giovanni Careri, *Bernini: Flights of Love, the Art of Devotion* (Chicago: The University of Chicago Press, 1995) 104.

226. Also of interest to *white noise viractualism* is the origin of the hermaphroditic image. This hybrid viractual image first appears in Ovid's classic text *Metamorphoses*, and perhaps this emergence is well worth recounting here. The hermaphrodite initially occurs in Western culture as a son of Hermes and Aphrodite named Hermaphroditus. Hermaphroditus was a typical, if exceptionally handsome, young male with whom the water nymph Salmacis fell madly in love. When Hermaphroditus rejected her sexual advances, Salmacis desired him even more. One spring day, Hermaphroditus stripped nude and dove into the pool of water that was Salmacis's habitat. Salmacis immediately dove in after him, embracing him and wrapping her body around his, just as, Ovid says, ivy does around a tree. She then prayed to the gods that she would never be separated from him, a prayer that they answered favorably. Consequently, Hermaphroditus emerged from the pool both man and woman. The patriarchal construction of woman as other and the female body as object is deeply rooted in the supposed duality (opposites) of the (two) sexes. Most feminist theory questions this patriarchal construction of sex and gender, suggesting that sex is expressed through a continuum, rather than as an opposing couplet based on heterosexist male/female polarities. Accordingly, within my viractual multiverse, containments designed for womanhood/manhood are subverted by the presence of ambiguous genitalia, the mutable image and performance of pan-sexuality. Gender here is viewed as an act of becoming. Consequently, gender performance fails to sustain sex oppression by ceasing to draw the boundaries of the Other. It is a provocation not only to male/female constructions of heterosexuality, but also to homosexual constructions of identity. This critique of "representation" in the aesthetic sense is part of a critique of "representation" in the political sense (and vice versa). Art, here, is seen as political in the sense that it is a site of power struggles that fail to presuppose a metaphysics which is itself a politics, one that establishes an order of values which often maintains the dominant order of meaning and power over breakthrough ideologies. As previously noted, the tale of Hermaphroditus suggests white noise viractualism is about pansexual eroticism melded to virtuality, quixotic transformation, and, of course, immersive noise-excess. The viractual realm is a political-spiritual chaosmos in the sense that new forms of sexual order arise such that any form of order is only temporary and provisional. But I don't think it is a chaosmos in the sense of

ceaseless flux and chaos. Rather, this sphere is attained through an emergent viractual operation, and I take abundant pleasure in imagining the forms of pan-order that arise within its algorithmic processes.

227. Giovanni Careri, *Bernini: Flights of Love, the Art of Devotion* (Chicago: The University of Chicago Press, 1995) 104.

228. Walter Stace, *The Teachings of the Mystics* (New York: New American, 1960) 11.

229. Curt von Westernhagen, *Wagner: A Biography Volumes 1 and 2* (London: Cambridge University Press, 1978) 332.

230. Jack Stein, *Richard Wagner and the Synthesis of the Arts* (Detroit: Wayne State University Press, 1960) 6.

231. Wassily Kandinsky, *The Art of Spiritual Harmony* (London: Constable, 1914) 5.

232. Wilfrid Blunt, *The Dream King: Ludwig II of Bavaria* (New York: Viking Press, 1970) 21-45.

233. Blunt, 143.

234. Michael Schuyt, Joost Elffers, and George Collins, *Fantastic Architecture: Personel and Eccentric Visions* (New York: Harry Abrams, 1980) 59.

235. Miller, 115-17.

236. Blunt, 234.

237. Miller, 116.

238. Blunt, 151.

239. Blunt, 151.

Chapter 4

240. *Chaos magic* is often highly individualistic and borrows liberally from other belief systems. In this way, some chaos magicians consider their practice to be an art-like metabelief and most chaos magicians routinely create magical symbols for themselves.

241. Joseph Nechvatal, *Immersive Ideals / Critical Distances* (Cologne: LAP Lambert Academic Publishing, 2009) 76-82.

242. Nechvatal, "The Artist and Familiars," *Blast* 1 (November/December, 1991).

243. For more on this, see my *Towards an Immersive Intelligence: Essays on the Work of Art in the Age of Computer Technology and Virtual Reality (1993-2006)* (New York: Edgewise Press, 2009).

244. Nechvatal, "The Artist and Familiars."

245. Joseph Nechvatal, and Didier Gagneur, eds. *Excess in the Techno-mediacratic Society* (Arbois: Musée d'Arbois, 1992).

246. For an in-depth focus of this work see: Matthew Biro, *The Dada Cyborg: Visions of the New Human in Weimar Berlin*, 65-103.

247. Werner Schmalenbach, *Kurt Schwitters* (New York: Abrams, 1970) 129-39.

248. Adrian Henri, *Total Art: Environments, Happenings, and Performance* (New York: Oxford University Press, 1974) 18-20.

249. Henri, 19.

250. Henri, 20.

251. Allan Kaprow, *Assemblage, Environments and Happenings* (New York: Abrams, 1966) 170-71.

252. Henri, 12.

253. In sound, this includes Filippo Tomaso Marinetti, Antonio and Luigi Russolo, Giacomo Balla, Rodolfo De Angelis, Carlo Belloli, Fortunato Depero and Francesco Cangiullo, among others.

254. They also sketched out a work called *Intuizione: Fiore magico transformabile motorumorista* (*Intuition: Magical transformable motor-noise flower*) in their futurist Manifesto *Ricostruzione dell'Universo* (1915).

255. Henri, 13-16.

256. Henri, 24.

257. Henri, 26.

258. Henri, 24.

259. The mechanamorphic impulse of Duchamp's works from 1911-1912, and the machine works that follow his exposure to Raymond Roussel, are an inescapable point of reference for the avant-garde of the 20th century. The machine in that century, for Duchamp, becomes the symbol of total bliss through pure mentality and auto-sexual autonomy in contradiction to the horror that mechanized war had brought. By hypnotizing our attention, the machine frees us from troubling obsessions and personal hang-ups through the alternative model of android life; intimating both our rush of desperation and our ecstatic release, refracted through a web of glazed impersonality. If the machine, as a representative of order, was a fascination Duchamp used to balance out the age's ineptness, whether of the mind or flesh, his mechanamorphic production and machine forms refigure the human body into an almost mechanized substance. In *The Bride Stripped Bare by the Bachelors, Even*, which positions a central bride machine over a bachelor apparatus, Duchamp, with the strictness of machinery, applies fantasy to seduction and masturbation. In a way, Duchamp suggests that we (as viewers) can use his art as a vehicle for self-transcendence into a kind of dream world of nonsense sex. By mechanizing sex and dreams, this nonsense of the sex machine converts sexual energy into artistic noise energy.

260. Henri, 22-4.

261. There had been a substantial eruption of this noise impetus following the war, as the reader shall soon see.

262. Herbert York, "Nuclear Deterrence and the Military Uses of Space," *Daedalus: Weapons in Space Vol. I: Concepts and Technologies* 114.2 (Spring 1985): 17-32; 20.

263. Raymond Roussel's repetitions, for example in his descriptions of eggs on plates and the multiple allusions to the odor of urine after the eating of asparagus, are typical of a poetic-mechanical apparatus helping to take us further into the area of the

unconscious and the noisy sexual. He points us towards an intellectual history that maps out art's role in creating social allegory, along with the mechanized mass killing of World War I and II, the holocaust, Hiroshima and the discovery of psychoanalysis (which is rooted in noise-sex symbolism) and so offers us an interesting context in which to view the possible role of the computer and noise art.

264. For a good overview of his work, see Elizabeth Frank, *Jackson Pollock* (New York: Abbeville Press, 1983).

265. See my "Immersive Implications" in Roy Ascott, ed., *Consciousness Reframed: Conference Proceedings* (Newport: CAiiA/University of Wales College, 1997).

266. Al Hansen, *A Primer of Happenings and Space/Time Art* (New York: Something Else Press, 1965) 6.

267. Henri, 162.

268. On March 9th, 1960, three nude female models painted each other with IKB Blue paint to the sounds of Klein's *Monotone Symphony* (which consisted of one note and an equally long silence, first written by Klein in 1949) and then gently pressed their bodies against the artistic ground.

269. Margot Lovejoy, *Postmodern Currents: Art and Artists in the Age of Electronic Media*. 2 nd Edition (Upper Saddle River: Prentice Hall, 1997) 55.

270. John Cage, *On Robert Rauschenberg, artist, and his work* (first published in *Metro, Milan*, 1961); republished in *Silence* 4th edition (M.I.T Press, Cambridge, Massachusetts, 1970) 13.

271. See my *An Ecstasy of Excess* (Mönchengladbach: Juni-Verlag, 1991).

272. Henri, 93.

273. Henri, 114.

274. Gene Youngblood, *Expanded Cinema* (New York: E. P. Dutton and Co, Inc., 1970) 368.

275. Andrew Solomon, "Dot Dot Dot," *Artforum* (February 1997): 66-73; 67.

276. Solomon, 70.

277. Margot Lovejoy, *Postmodern Currents: Art and Artists in the Age of Electronic Media: Second Edition* (Upper Saddle River: Prentice Hall, 1997) 56.

278. Solomon, 67.

279. See my "Immersive Implications," *New Observations* 116 (Fall 1997): 46-7.

280. Bill Morgan, *I Celebrate Myself: The Somewhat Private Life of Allen Ginsberg* (New York: Penguin Books, 2007) 318.

281. Burroughs lusted after Ginsberg in vain.

282. Morgan, 318-19.

283. Morgan, 331-32.

284. Only by being really difficult can the child discover whether the parent is resilient and robust. In like fashion, noise art must be difficult—or we will never find out what the world (and art) are really like.

285. From the early 1990s.

286. See Tiziana Terranova, *Network Culture: Politics for the Information Age* (London: Pluto Press, 2004).

287. See Frank Popper, *Art, Action and Participation* (New York: New York University, 1975). See also his *From Technological to Virtual Art* (Cambridge: MIT Press, 2007).

288. Rebuilt in 1970 and now in the collection of Harvard University's Busch-Reisinger Museum.

289. Stan Vanderbeek, "Movies: Disposable Art, Synthetic Media and Artificial Intelligence." *Take One Film Magazine* 2.3 (1969): 14-16; 16.

290. Stan Vanderbeek, "Culture: Intercom and Expanded Cinema: A Proposal and Manifesto." *Film Culture* 40 (Spring, 1966): 15-18; 16.

291. Vanderbeek, "Culture: Intercom and Expanded Cinema: A Proposal and Manifesto," 17.

292. Jean-Christophe Royoux, "Expanded, Extended: Héritage, Transformation et Ramifications d'un Concept Esthétique dans l'Art Années Soixante," *Omnibus* 23 (Janvier, 1998): 7.

293. Youngblood, 354-58.

294. Youngblood, 66.

295. Andy Warhol and Pat Hackett, *POPism: The Warhol' 60*. (London: Hutchinson, 1980) 156.

296. Experiments in Art and Technology (E.A.T.) was a non-profit and tax-exempt organization established to develop collaborations between artists and engineers. E.A.T. initiated and carried out projects that expanded the role of the artist in contemporary society and helped eliminate the separation of the individual from technological change. E.A.T. was never a concrete channel that formalized an art-science interchange in some elaborate bureaucratic institution. Rather it served to facilitate person-to-person contacts between artists and engineers. It was officially launched in 1967 by the engineers Billy Klüver and Fred Waldhauer with the artists Robert Rauschenberg and Robert Whitman. These men had previously collaborated, most notably in 1966 when they together organized *9 Evenings: Theater and Engineering*, a series of performance art presentations that united artists and engineers. The performances were held in New York City's 69th Regiment Armory, on Lexington Avenue between 25th and 26th Streets as an homage to the original and historical 1913 Armory show. Such collaborations continued to break down barriers between the arts and scientists in the 1960s, 1970s, and 1980s and indirectly launched and supported the experimental sound artist John Cage, dancer Merce Cunningham, and pop artist Andy Warhol. The pinnacle of E.A.T. activity is generally considered to be the Pepsi Pavilion at Expo '70 at Osaka Japan where E.A.T. artists and engineers collaborated to design and program an immersive dome.

297. Youngblood, 416-17.

298. Solomon, "Dot Dot Dot," 67.

299. Solomon, "Dot Dot Dot," 68.

300. Jacques Lacan, "The Mirror Stage as Formative of the Function of the I," *Ecrits: A Selection* (New York: Norton) 1-7.

301. Janet Wolff, *The Social Production of Art* (New York: New York University Press, 1993) 132-36.

302. Popper, 158.

303. Popper, 92.

304. Youngblood, 359.

305. Martin called himself ST EOM.

306. Youngblood, 371.

307. Youngblood, 391-92.

308. Youngblood, 387-91.

309. Youngblood, 389.

310. Youngblood, 381-83.

311. For more on soundscapes, see R. Murray Schafer, *The Soundscape: Our Sonic Environment and the Turning of the World* (Rochester, VT: Destiny Books, 1984).

Chapter 5

312. For more on net art viruses, see JussiParikka, "Archives of Software: Computer Viruses and the Aesthesis of Media Accidents," in *The Spam Book: On Viruses, Porn, and Other Anomalies from the Dark Side of Digital Culture*. Eds. Jussi Parikka, and Tony Sampson (Cresskill: Hampton Press, 2009) 105-24.

313. Examples of viral net art are *biennale.py*, the computer virus exhibited at the 2001 Venice Biennale by 0100101110101101.org, and much of the work of Jodi, a collaboration that uses the dysfunctions and the potential break down of network software as artistic potential.

314. Knowbotic Research KR+cF has regularly invited people from non-art fields to participate in their projects, such as scientists, philosophers and engineers, depending on the concept of each project.

315. One here might recall Luigi Russolo's idea of *Rete dei rumori* (*Network of Noises*) published in 1914 in the magazine *Lacerba*.

316. See for example http://assembler.org/axbx/ax.html and/or http://meta.am/

317. See Roy Ascott, *Telematic Embrace, Visionary Theories of Art, Technology and Consciousness* (Berkeley: University of California Press, 2003).

318. A fine example of this overriding is the aforementioned Stan Vanderbeek's 1966 proposal essay "Culture: Intercom and Expanded Cinema: A Proposal and Manifesto," that was published in *Film Culture* 40 (Spring, 1966). In this essay on page 17, Vanderbeek called for the transformation of his Movie Dromes into "image libraries" which by "computer inter-play" would function as global "communication and storage centers". According to Vanderbeek, "by satellite, each dome could receive its images from a world wide library source, store them and program feedback

presentations to the local community". Vanderbeek also went so far as to predict that such a linking of visual "feedback" could "authentically review the total world image reality" and hence produce "a sense of the entire world picture". This process of linking visual dome-worlds Vanderbeek labelled "intra-communitronics".

319. Jussi Parikka, *Digital Contagions: A Media Archaeology of Computer Viruses* (New York: Peter Lang, 2007) 136.

320. The word *parasite*, interestingly enough, also means *noise* in French. For more on this curious alliance, see Michel Serres's book *The Parasite* (Paris: Grasset, 1980); *The Parasite*, trans. Lawrence R. Schehr (Baltimore: Johns Hopkins University Press, 1982) and *Reading with Michel Serres* by Maria L. Assad (Albany: SUNY Press, 1999).

321. Briefly, the *viractual* is the stratum of activity where distinct actualizations/individuations are materialized out of the flow of virtuality.

322. Parikka, 5.

323. This reminds me of the closing statement of Tristan Tzara in his "Dada Manifesto" where he states, "Perhaps you will understand me better when I tell you that Dada is a virgin microbe that penetrates with the insistence of air into all the spaces that reason has not been able to fill with words or conventions". From "Dada Manifesto" [1918] and "Lecture on Dada" [1922], translated from the French by Robert Motherwell in *The Dada Painters and Poets: An Anthology* 2nd Edition, (Cambridge: Harvard University Press, 1989) 78-9.

324. Retroviruses are sometimes known as anti-anti-viruses. The basic principle is that the virus must somehow hinder the operation of an anti-virus program in such a way that the virus itself benefits from it. Anti-anti-viruses should not be confused with anti-virus-viruses, which are viruses that will disable or disinfect other viruses.

325. Heuristic virus cleaners work by loading an infected file up to memory and emulating the program code. It uses a combination of disassembly, emulation and sometimes execution to trace the flow of the virus and to emulate what the virus is normally doing. The risk in heuristic cleaning is that if the cleaner tries to emulate everything, the virus might get control inside the emulated environment and escape, after which it can propagate further or trigger a destructive retaliation reflex.

326. See http://www.henryflynt.org/and Henry Flynt, "Mutations of the Vanguard: Pre-Fluxus, During Fluxus, Late Fluxus," in A.B. Olivia, *Ubi Fluxus ibi motus 1990—1962* (Milano: Mazzotta, 1990) 99-128.

327. Such thinking relates to my 2001 *Computer Virus Project 2.0* that followed along the same lines as my previous viral works by in 1992—works where an unpredictable progressive virus operates on a degradation/transformation of an image. Using a C++ framework, I and my programmer and collaborator Stephane Sikora brought my early computer virus project into the realm of artificial life (A-Life) (i.e. into a synthetic system that exhibits behaviors characteristic of natural living systems). With *Computer Virus Project 2.0*, elements of artificial life have been introduced in that viruses are modeled to be autonomous agents living in/off the image. The project simulates a population of active viruses functioning as an analogy of a viral biological system. *Computer Virus Project 2.0* actively propagates viral attacks on image-files from my "ec-satyricOn 2000 (enhanced)+ bodies in the bit-stream (compliant)" series. Here viral algorithms—based on a viral biological model—are used to define evolutionary processes which are then applied to image-files from my *ec-satyricOn 2000 (enhanced)+ bodies in the bit-stream (compliant)* series. Among the different

techniques used here were models that result from embodied artificial intelligence and the paradigm of genetic programming. The world is modeled as an image via a set of pixels. Every pixel's color is defined by RGB real number vectors that represent the red, green and blue components of every pixel's color. The image world has no edges. Every square on the edge of the image is adjacent to another on the opposite edge. A virus can move around the image and impact the image world as different colors actually correspond to resources used for survival by the viruses. The behavior of a virus is modeled as a generated looping activity that is typical of situated artificial intelligence work. A virus will pick up information from its environment, decide on a course of action, and carry it out. The loop is simplified here because of the abstract character of the simulacrum. Viral instructions provide different possibilities for executing instructions according to the environmental conditions in which the virus is living. A virus will perceive the pixel it is on and the eight adjacent ones. It can get information on its color and on the possible presence of other viruses. In order to decide on a course of action, each virus is programmed with a set of randomized instructions of different kinds; some relate to direction, others to a change in the color of the current pixel (the one the virus is in). Others control the implementation of the program and carry out tests. Once the program has been executed, following actions to be carried out randomly arise. As the virus executes them, it moves to one of the adjacent squares and changes the current pixel. It can even reproduce itself (reproduction here results from the instruction 'divide'). A virus that carries out that instruction will produce a replica of itself—although slightly altered. It's genome-program changes with the mutation operator. In addition to these changes, every cycle produces a change in the energy level of the virus. The virus will lose a set amount of energy with every run, and when it runs out of energy, it dies (i.e. it disappears). In order to survive, a virus needs to pick up energy, which it can only do by degrading the image. The more it changes the color of a pixel, the more energy it acquires. The difference between the color before and after is calculated. We can see from a virus's behavior and direction whether it will be more or less adaptable—more or less able to survive. There are a maximum number of viruses that can be present simultaneously (usually 1000). When that number is reached, the 'divide' instruction is ignored. If the virus has enough energy it will move around randomly, otherwise it will follow its favorite color and absorb part of the red component of the pixel it is on.

328. It is amusing to recall here that Pierre Jaquet-Droz created the first robotic mechanical figure in 1774 called the *automatic scribe*. It still can be seen at the Musee d'Art et Histoire in Neuchatel, Switzerland.

Conclusion

329. David Chalmers, "Facing up to the Problems of Consciousness," *Journal of Consciousness Studies: Controversies in Science and the Humanities*, 2.3 (1995): 200-219; 211.

330. Gilles Deleuze and Félix Guattari, *What is Philosophy?* (London: Verso Books, 1994) 132.

331. Chalmers, 211.

332. A rhizome literally is a root-like plant stem that forms a large entwined spherical zone of small roots which criss-cross. In the philosophical writings of Deleuze and Guattari the term is used as a metaphor for an epistemology (that in philosophy which is concerned with theories of knowledge) that spreads in all directions simultaneously.

(Deleuze and Guattari, *What is Philosophy?* 7.) More specifically, Deleuze and Guattari define the rhizome as that which is "reducible to neither the One or the multiple. It has neither beginning nor end, but always a middle (milieu) from which it grows and which it overspills. It constitutes linear multiplicities with n dimensions having neither subject nor object... ". Deleuze and Guattari, *A Thousand Plateaus: Capitalism and Schizophrenia*. (Minneapolis: University of Minnesota Press, 1987) 21.

333. Visionary art is art that purports to transcend the physical world and portray a wider vision of awareness including spiritual, ecstatic or mystical themes—or is based in such experiences—accessible through the subjective realm of each individual. What unites visionary artists is the driving force and source of their art: their unconventionally intense psychic imaginations. Their gift to the world is to reveal "in minute particulars" (as William Blake would say) the full spectrum of the vast visionary dimensions of the mind. William Blake, for example, is famous for his identifying the entirety of the universe in a single grain of sand. Both trained and self-taught (*Art Brut* or *Outsider Art*) artists have, and continue to create visionary works. The famous fantastical and visionary 15th century painter Hieronymous Bosch is a good example of the trained sort. Also important is French Symbolism (Gustav Moreau and Odilon Redon) and Dada's use of chance automatic irrational procedures (which grew into Surrealist activities of Max Ernst, Salvador Dali, Hans Arp, Hans Bellmer and Juan Miro). The visions of the Surrealists help to define a dream realm where bizarre juxtaposition is possible and desirable. A profound truth resides in such strangeness, for these visions can shock us into deepening our acknowledgement and appreciation of the great mystery of the universe.

334. Achieved through re-establishing a sense of self wonder through forms of noise where corporations and governmental agencies are not welcome.

335. The milieu of sound/image superabundance and proliferation-collapse.

336. Noise art may be construed as a personal technique for taking back a bit of the raw actual world from the slick media world.

337. This quality has formulated a new understanding of phallocratic existence which Deleuze and Guattari have called *schizoid*. According to them, being is now inseparable from a technologically hallucinogenic/schizoid culture. Deleuze, and Guattari, *A Thousand Plateaus: Capitalism and Schizophrenia* (Minneapolis: University of Minnesota Press, 1987).

338. Deleuze and Guattari, *On The Line* (New York: Semiotext(e), 1983) 2.

339. Paul Hegarty, *Noise Music: A History* (London: Continuum, 2007) 122.

340. Perhaps noise is the trickster/joker necessary to the health of a system, as dysfunctioning remains essential for functioning.

341. Alfred Jarry, "What is Pataphysics?" *Evergreen Review* 13 (1963): 131-151; 131.

342. The concept of *Maya* in Indian philosophy refers to the purely phenomenal, insubstantial character of the everyday world.

343. Perhaps it is relevant here to remember that Mannerism (generally the art of the period of Late-Renaissance circa 1530-1600) was an aesthetic movement that valued highly refined gracefulness and elegance: a beautiful *maniera* (style) from which Mannerism takes its name. The term usually means an art in which lavish attention is paid to stylization and to the superficialities of semblance.

344. Deleuze and Guattari, *Nomadology: The War Machine* (New York: Semiotext(e), 1986) 13.

345. Deleuze and Guattari, *Nomadology*, 36.

346. John Cage has written, "It becomes evident that music itself is an ideal situation, not a real one. The mind may be used either to ignore ambient sounds, pitches other than the eighty-eight, durations which are not counted, timbres which are unmusical or distasteful, and in general to control and understand an available experience. Or the mind may give up its desire to improve on creation and function as a faithful receiver of experience". John Cage, *Silence* (London: Calder and Boyers, 1961) 32.

347. This corresponds to Douglas Kahn's contention that noise drifts across the binary empirical/abstract, such that "when noise itself is being communicated, [...] it no longer remains inextricably locked into empiricism but is transformed into an abstraction of another noise". Douglas Kahn, *Noise, Water, Meat: A History of Sound in the Arts* (Cambridge: MIT Press, 2001) 25.

348. Gilles Deleuze, *Spinoza: Practical Philosophy* (San Francisco: City Lights, 1984) 21.

349. Non-narrowly empiricist.

350. A loss of cognitive body-image consistent with Arthur Schopenhauer's (1788-1860) conception of a *pure knowing subject*. Arthur Schopenhauer, *The World as Will and Idea* (London: Kegan Paul, Trench, Truber and Co., 1907) 127.

351. See my small book *An Ecstasy of Excess* (Mönchengladbach: Juni-Verlag) 3-7.

352. Kendall Walton, *Mimesis as Make-Believe: On the Foundations of the Representational Arts* (London: Harvard University Press, 1990) 11.

353. Friedrich Nietzsche, *Beyond Good and Evil* (Edinburgh: Darien Press, 1907) 7.

354. Robert Romanyshyn, *Technology as Symptom and Dream* (London: Routledge, 1989) 83-93.

355. Hal Foster, ed. *Vision and Visuality* (Seattle: Bay Press, 1988) x.

356. The basic neurologically informed concepts of existence.

357. Vladimir Nabokov, *Pale Fire* (New York: Random House, 1989) 69.

358. Ronald Bogue, "Word, Image and Sound: The Non-Representational Semiotics of Gilles Deleuze," in Ronald Bogue, ed. *Mimesis, Semiosis and Power* (Philadelphia: John Benjamins, 1991) 83-4.

359. Wolfgang Iser, *The Act of Reading: A Theory of Aesthetic Responses* (London: Routledge and Kegan Paul, 1978) 21.

360. I remind the reader here that the emancipation of noise in music via John Cage is not the exclusive responsibility of the composer or the musician, but requires an active and transformable attitude on the listener's part as well. When the listener includes the sounds of the environment in a musical composition, (s)he in fact becomes co-composer.

361. Brian Massumi, "The Autonomy of Affect," *Cultural Critique* (Fall 1995): 83-109; 96. In this regard, I suggest that readers listen to the *Big Bang Sound*, a noise simulation of the first (and last) sound as derived from the sound propagating as compression waves through the plasma/hydrogen medium of the early universe some 100 to 700

thousand years after the initial Big Bang at http://www.astro.virginia.edu/~dmw8f/sounds/cdromfiles/index.php (accessed 11/11/2008)

362. Jean-François Lyotard, "Que Peindre?: Interview with Bernard Macade," *Art Press* 125 (May 1988): 42-5; 45.

363. Jacques Derrida, "The Theater of Cruelty and the Closure of Representation," in Jacques Derrida, *Writing and Difference* (Chicago: University of Chicago Press, 1978) 232-250; 235.

364. See my "Introduction to The Art of Excess in the Techno-mediacratic Society," *New Observations* 94 (1993).

365. Dick Higgins, *A Dialectic of Centuries: Notes Towards a Theory of the New Arts* (New York: Printed Editions, 1978) 12-17.

366. Higgins, 3-9.

367. Henry Flynt. "Mutations of the Vanguard: Pre-Fluxus, During Fluxus, Late Fluxus," A. B. Olivia, 107.

368. Higgins, 8.

369. For more on this topic, see my book *Towards an Immersive Intelligence: Essays on the Work of Art in the Age of Computer Technology and Virtual Reality (1993-2006)* (New York: Edgewise Press, 2009).

370. Quoted in Carla Hoekendijk, Ed. *Interfacing Realities* (Amsterdam: V2 Organisatie, 1997) 14.

371. Gilles Deleuze, *Logic of Sense* (New York: Columbia University Press, 1990) 151.

372. Brian Massumi, "The Autonomy of Affect," *Cultural Critique* (Fall 1995): 83-109; 91.

373. Though I have never seen it, apparently a satisfactory example of superimposed cinema was the December 1967 screening of Andy Warhol's **** *(FOUR STARS)* 16mm/25 hrs/sound/color/24 fps that was filmed during August 1966 to September 1967. Two projectors' images were superimposed on a single screen, mixing both films (color and b&w) as well as the two soundtracks together.

374. Gene Youngblood, *Expanded Cinema* (New York: E. P. Dutton and Co, Inc., 1970) 111.

375. Youngblood, 87.

376. One creating movement/pertubation in societies that would otherwise rest stagnate.

377. Giovanni Careri, *Bernini: Flights of Love, the Art of Devotion* (Chicago: The University of Chicago Press, 1995) 104.

378. Jacques Derrida, "The Theater of Cruelty and the Closure of Representation," in Jacques Derrida, *Writing and Difference* (Chicago: University of Chicago Press, 1978) 232-50.

379. For example, there is part of ourselves that has little concern with our best interests.

380. Viruses, by biological definition, must be considered as semi-live.

381. For example, see Leon Cohen's paper, "The History of Noise" in which Cohen makes the case for noise being involved in the resolution of major scientific, mathematical, and technological problems.

382. A noise art exception here is datamoshing, the manipulating of digital compression to produce pixel bleeding for artistic effect.

383. Excluding the prior examples of noise film cited in this text along with these Dada films: Viking Eggeling's *Diagonal Symphony*, 1921; Paul Strand/Charles Sheeler's *Manhattan*, 1921; Hans Richter's *Rhythmus 21*, 1921, *Rhythmus 23*, 1923-1925, *Filmstudie*, 1926, and the *Vormittagsspuk*, 1927-1928; René Clair's *Entr'acte*, 1924; and Marcel Duchamp's *Anémic Cinema*, 1925.

384. Albert Rothenberg, *The Emerging Goddess* (Chicago: University of Chicago, 1979) 345.

385. Rothenberg, 12.

386. Regardless of Luigi Russolo's participation in the pro-fascist (pro-war) Futurist movement, one must remember that actual right-wing fascist dictators have little use for avant-garde noise art. They much prefer folk art and popular music.

387. Paul Hegarty, *Noise Music: A History* (London: Continuum, 2007) 125.

388. This glut of frictionlessness flow was exemplified by the fluid movement (leading to collapse) of money markets for mortgage-backed securities and derivatives unhinged to tangible value: where the meaninglessness of huge abstract numbers is slickly numbing.

389. In this sense, noise art equates to the sound of the rage of the sea—the sea being the source of all life.

390. For a probing investigation of this subject, see Quentin Meillassoux's *After Finitude: An Essay on the Necessity of Contingency* (London: Continuum, 2008).

Bibliography

Adorno, Theodor. *Aesthetic Theory*. London: Routledge and Kegan Paul. 1984.

Ascott, Roy. *Telematic Embrace, Visionary Theories of Art, Technology and Consciousness*. Berkeley: University of California Press. 2003.

Assad, Marina L. *Reading with Michel Serres*. Albany: SUNY Press. 1999.

Attali, Jacques. *Noise: The Political Economy of Music*. Minneapolis: University of Minnesota Press. 1985.

Bailey, Thomas Bey William. *Micro Bionic: Radical Electronic Music And Sound Art In The 21st Century*. Creation Books. 2009.

Bataille, Georges. *Oeuvres Completes: Lascaux: La Naissance de l'Art*. Paris: Gallimard. 1979.

Bataille, Georges. *Visions of Excess*. Minneapolis: University of Minnesota Press. 1985.

Berger, John. *Ways of Seeing*. New York: Penguin Books. 1977.

Bersani, Jacques, et al. eds. *Archéologie: Les Grands Atlas Universalis*. Paris: Encyclopedia Universalis. 1985.

Biro, Mathew. *The Dada Cyborg: Visions of the New Human in Weimar Berlin*. Minneapolis: University of Minnesota Press. 2009.

Blunt, Wilfrid. *The Dream King: Ludwig II of Bavaria*. New York: Viking Press. 1970.

Bogue, Ronald. "Word, Image and Sound: The Non-Representational Semiotics of Gilles Deleuze". Bogue, Ronald. ed. *Mimesis, Semiosis and Power*. Philadelphia: John Benjamins, 1991.

Brassier, Ray. "Genre is Obsolete". *Multitudes*, No. 28, Spring 2007.

Brennan, Teresa and Jay, Martin. eds. *Vision in Context*. London: Routledge. 1996.

Cage, John. *Silence*. London: Calder and Boyers. 1961.

Careri, Giovanni. *Bernini: Flights of Love, the Art of Devotion*. Chicago: The University of Chicago Press. 1995.

Cascone, Kim. "The Aesthetics of Failure: 'Post-Digital' Tendencies in Contemporary Computer Music". *Computer Music Journal* 24, no. 4. Winter. 2002. pp. 12-18.

Chalmers, David. "Facing up to the Problems of Consciousness". *Journal of Consciousness Studies: Controversies in Science and the Humanities*, Vol. 2, No. 3, 1995. pp. 200-219.

Cohen. Leon. "The History of Noise on the 100th Anniversary of its Birth". *IEEE Signal Processing Magazine*. November. 2005.

Collingwood, R. G. *Principles of Art*. Oxford: Claredon Press. 1938.

Croix, Horst de la and Tansey, Richard. *Gardner's Art through the Ages*. New York: Harcourt, Brace and Javanovich. 1975.

Crowther, Paul. *Critical Aesthetics and Postmodernism*. Oxford: Clarendon Press. 1993.

Debord, Guy. *The Society of the Spectacle*. Detroit: Black and Red. 1976.

Deleuze, Gilles and Guattari, Félix. *A Thousand Plateaus: Capitalism and Schizophrenia*. Minneapolis: University of Minnesota Press. 1987.

Deleuze, Gilles and Guattari, Félix. *Nomadology: The War Machine*. New York: Semiotext(e). 1986.

Deleuze, Gilles and Guattari, Félix. *On The Line*. New York: Semiotext(e). 1983.

Deleuze, Gilles and Guattari, Félix. *What is Philosophy?*. London: Verso Books. 1994.

Deleuze, Gilles. "Postscript on Control Societies". *Negotiations: 1972-1990*. New York: Columbia. 1995.

Deleuze, Gilles. *Logic of Sense*. New York: Columbia University Press. 1990.

Deleuze, Gilles. *Spinoza: Practical Philosophy*. San Francisco: City Lights. 1984.

Delluc, Brigitte and Delluc, Gilles. *Discovering Lascaux*. Pollina à Luçon: Editions Sud Ouest. 1990.

Demers, Joanna. *Listening Through The Noise*. New York: Oxford University Press. 2010.

Derrida, Jacques. "The Theater of Cruelty and the Closure of Representation".

Derrida, Jacques. *Writing and Difference*. Chicago: University of Chicago Press. 1978. pp. 232-250.

Edgerton, Samuel. *The Renaissance Discovery of Linear Perspective*. New York: Harper and Row. 1976.

Ehrenzweig, Anton. *The Hidden Order of Art*. Berkeley: University of California Press. 1967.

Flynt, Henry. "Mutations of the Vanguard: Pre-Fluxus, During Fluxus, Late Fluxus". Olivia, Achille Bonito. *Ubi Fluxus ibi motus 1990-1962*. Milan: Mazzotta. 1990. pp. 99-128.

Foster, Hal, ed. *Vision and Visuality*. Seattle: Bay Press. 1988.

Foucault, Michel. *Discipline and Punish: The Birth of the Prison*. New York: Vintage. 1979.

Frank, Elizabeth. *Jackson Pollock*. New York: Abbeville Press. 1983.

Goodman, Steve. "Contagious Noise: From Digital Glitches to Audio Viruses". Parikka, Jussi and Sampson, Tony D. eds. *The Spam Book:*

On Viruses, Porn and Other Anomalies From the Dark Side of Digital Culture. Cresskill: Hampton Press. 2009.

Goodman, Steve. *Sonic Warfare: Sound, Affect, and the Ecology of Fear*. Cambridge: MIT Press. 2010.

Hansen, Al. *A Primer of Happenings and Space/Time Art*. New York: Something Else Press. 1965.

Harrison, Thomas. J. *The Emancipation of Dissonance*. Berkeley: University of California Press. 1910.

Harrison, Jane Ellen. *Ancient Art and Ritual*. Bradford-on-Avon, Wilts: Moonraker Press. 1913.

Hegarty, Paul. *Noise Music: A History*. London: Continuum. 2007.

Henri, Adrian. *Total Art: Environments, Happenings, and Performance*. New York: Oxford University Press. 1974.

Higgins, Dick. *A Dialectic of Centuries: Notes Towards a Theory of the New Arts*. New York: Printed Editions. 1978.

Hoekendijk, Carla, ed. *Interfacing Realities*. Amsterdam: V2 Organisatie. 1997.

Iles, Anthony and Iles, Mattin eds. *Noise & Capitalism*. Donostia-San Sebastián: Arteleku Audiolab. Kritika series. 2009.

Iser, Wolfgang. *The Act of Reading: A Theory of Aesthetic Responses*. London: Routledge and Kegan Paul. 1978.

Ivins, William. *Art and Geometry*. New York: Dover. 1964.

Jarry, Alfred. "What is Pataphysics?". *Evergreen Review*, No. 13. 1963. pp. 131-151

Johnston, John. *The Allure of Machinic Life: Cybernetics, Artificial Life, and the New AI*. Cambridge: MIT Press. 2008.

Kahn, Douglas. *Noise, Water, Meat: A History of Sound in the Arts*. Cambridge: MIT Press. 2001.

Kandinsky, Wassily. *The Art of Spiritual Harmony*. London: Constable and Co. Ltd. 1914.

Kaprow, Allan. *Assemblage, Environments and Happenings*. New York: Abrams. 1966.

Kelly, Caleb. *Cracked Media: The Sound of Malfunction*. Cambridge: MIT Press. 2009.

Kemp, Martin. *The Science of Art: Optical Themes in Western Art from Brunelleschi to Seurat*. New Haven: Yale University Press. 1990.

Kernberg, Otto. *Borderline Conditions and Pathological Narcissism*. New York: Jason Aronson. 1975.

Kinsey, Alfred. *Sexual Behavior in the Human Female*. Philadelphia: Saunders. 1953.

Koestler, Arthur, ed. *Beyond Reductionism*. Boston: Beacon. 1971.

Koestler, Arthur. *The Ghost in the Machine*. New York: Hutchinson, London and Macmillan. 1967.

LaBelle, Brandon. *Background Noise: Perspectives on Sound Art*. London: Continuum. 2008.

Lacan, Jacques. "The Mirror Stage as Formative of the Function of the I". Lacan, Jacques. *Ecrits: A Selection*. New York: Norton. 1977. pp. 1-7.

Langer, Susanne. *Feeling and Form: A Theory of Art Developed from Philosophy in a New Key*. New York: Charles Scribner's Sons. 1953.

Leroi-Gourhan, André. *The Art of Prehistoric Man in Western Europe*. London: Thames and Hudson. 1968.

Levinas, Emmanuel. *Totality and Infinity: An Essay on Exteriority*. Pittsburgh: Duquesne University Press. 1969.

Lilly, John C. *Programming and Metaprogramming in the Human Bio-Computer: Theory and Experiments*. New York: Bantam Books. 1974.

Lovejoy, Margot. *Postmodern Currents: Art and Artists in the Age of Electronic Media*. 2nd Edition. Upper Saddle River: Prentice Hall. 1997.

Lowe, Donald. *History of Bourgeois Perception*. Chicago: University of Chicago. 1982.

Lyotard, Jean-François. "Que Peindre?: Interview with Bernard Macade". *Art Press* 125, May 1988, pp. 42-45

Massumi, Brian. "The Autonomy of Affect". *Cultural Critique*, Fall Issue. 1995. pp. 83-109.

Massumi, Brian. *A User's Guide to Capitalism and Schizophrenia: Deviations from Deleuze and Guattari*. Cambridge: MIT Press. 1992.

Meillassoux, Quentin. *After Finitude: An Essay on the Necessity of Contingency*. London: Continuum. 2008.

Metzinger, Thomas. ed. *Conscious Experience*. Paderborn: Schöningh. 1995.

Miller, Naomi. *Heavenly Caves: Reflections on the Garden Grotto*. London: Allen and Unwin. 1982.

Mithen, Steven. *The Prehistory of the Mind: The Cognitive Origins of Art, Religion and Science*. Cambridge: Cambridge University Press. 1996.

Morgan, Bill. *I Celebrate Myself: The Somewhat Private Life of Allen Ginsberg*. New York: Penguin Books. 2007.

Motherwell, Robert. *The Dada Painters and Poets: An Anthology*. 2nd Edition. Cambridge: Harvard University Press. 1989.

Nabokov, Vladimir. *Pale Fire*. New York: Random House. 1989.

Nechvatal, Joseph. *Immersive Ideals / Critical Distances*. Cologne: LAP Lambert Academic Publishing. 2009.

Nechvatal, Joseph. *Towards an Immersive Intelligence: Essays on the Work of Art in the Age of Computer Technology and Virtual Reality (1993-2006)*. New York: Edgewise Press. 2009.

Nechvatal, Joseph. "An Ecstasy of Excess". Nechvatal, Joseph. *An Ecstasy of Excess*. Mönchengladbach: Juni-Verlag. 1991. pp. 3-7.

Nechvatal, Joseph. "Hyper-death and the Scopic Corpse". *Artforum*. November Issue. 1990. p. 130.

Nechvatal, Joseph. "Immersive Implications". Ascott, Roy. ed. *Consciousness Reframed: Conference Proceedings*. Newport: CAiiA/ University of Wales College. 1997.

Nechvatal, Joseph. "Immersive Implications". *New Observations*, No. 116, Fall. 1997. pp. 46-47.

Nechvatal, Joseph. "Introduction to The Art of Excess in the Techno-mediacratic Society". *New Observations*, No. 94. 1993.

Nechvatal, Joseph. "New Territory of Significance". *New Observations*, No. 104. 1994.

Nechvatal, Joseph. "The Art of Excess in the Techno-mediacratic Society". *New Observations*, No. 94. 1993.

Nechvatal, Joseph. "The Artist and Familiars". *Blast*, Vol. 1, November / December. 1991.

Nechvatal, Joseph. *An Ecstasy of Excess*. Mönchengladbach: Juni-Verlag. 1991.

Nechvatal, Joseph, and Gagneur, Didier. eds. *Excess in the Techno-mediacratic Society*. Arbois: Musée d'Arbois. 1992.

Nechvatal, Joseph. *Selected Writings*. Paris: Editions Antoine Candau. 1990.

Nietzsche, Friedrich. *Beyond Good and Evil*. Edinburgh: Darien Press. 1907.

O'Neil, Joseph, ed. *Mexico: Splendors of Thirty Centuries*. New York: Metropolitan Museum of Art. 1990.

Ostwald, Madelene. "Structuring Virtual Urban Space: Aboroscent Schemes". Droege, Peter, ed. *Intelligent Environments: Spatial Aspects of the Information Revolution*. Amsterdam: Elsevier. 1997. pp. 451- 482.

Panofsky, Erwin. *Tomb Sculpture: Four Lectures on its Changing Aspects from Ancient Egypt to Bernini*. New York: Abrams. 1991.

Parikka, Jussi. "Archives of Software: Computer Viruses and the Aesthesis of Media Accidents". *The Spam Book: On Viruses, Porn, and Other Anomalies from the Dark Side of Digital Culture*. eds. Parikka, Jussi and Sampson, Tony. Cresskill: Hampton Press. 2009.

Parikka, Jussi. *Digital Contagions: A Media Archaeology of Computer Viruses*. New York: Peter Lang. 2007.

Pennick, Nigel. *The Ancient Science of Geomancy*. London: Thames and Hudson. 1979.

Popper, Frank. *From Technological to Virtual Art*. Cambridge: MIT Press. 2007.

Popper, Frank. *Art—Action and Participation*. New York: New York University. 1975.

Porphyry. *L'Antro des Nymphes*. Paris: Pléiade. 1918.

Pylyshyn, Zenon. "Here and There In the Visual Field". Pylyshyn, Zenon. ed. *Computational Processes in Human Vision: An Interdisciplinary Perspective*. Norwood: Ablex. 1988. pp. 210-238.

Rheingold, Howard. *Virtual Reality*. New York: Summit Books. 1991.

Romanyshyn, Robert. *Technology as Symptom and Dream*. London: Routledge. 1989.

Rose, Jacqueline. *Sexuality in the Field of Vision*. London: Verso. 1986.

Rothenberg, Albert. *The Emerging Goddess*. Chicago: University of Chicago. 1979.

Roussel, Raymond. *How I Wrote Certain of My Books*. Cambridge: Exact Change. 2005.

Royoux, Jean-Christophe. "Expanded, Extended: Héritage, Transformation et Ramifications d'un Concept Esthétique dans l'Art Annees Soixante". *Omnibus n°23* January. 1998. pp. 5-8.

Rudgley, Richard. *The Alchemy of Culture: Intoxicants in Society*. London: British Museum Press. 1993.

Ruspoli, Mario. *The Cave of Lascaux: The Final Photographic Record*. New York: Abrams. 1987.

Russolo, Luigi. *The Art of Noises*. New York: Pendragon. 1986.

Schafer, R. Murray. *The Soundscape: Our Sonic Environment and the Turning of the World*. Rochester: Destiny Books. 1984.

Schmalenbach, Werner. *Kurt Schwitters*. New York: Abrams. 1970.

Schopenhauer, Arthur. *The World as Will and Idea*. London: Kegan Paul, Trench, Truber and Co. 1907.

Schuyt, Michael, Elffers, Joost, and Collins, George. *Fantastic Architecture: Personel and Eccentric Visions*. New York: Harry Abrams. 1980.

Serres, Michel. *Genèse*. Paris: B. Grasset. 1982.

Serres, Michel. *The Parasite*. Minneapolis: University of Minnesota Press. 2007.

Solomon, Andrew. "Dot Dot Dot". *Artforum*, February Issue. 1997. pp. 66-73.

Stace, Walter. *The Teachings of the Mystics*. New York: New American. 1960.

Stein, Jack. *Richard Wagner and the Synthesis of the Arts*. Detroit: Wayne State University Press. 1960.

Stewart, Robert, ed. *Ideas That Shaped Our World: Understanding the Great Concepts of Then and Now*. London: Marshall. 1997.

Talbott, Stephen. *The Future Does Not Compute: Transcending the Machines in Our Midst*. Sebastopol: O'Reilly and Associates. 1995.

Terranova, Tiziana. *Network Culture: Politics for the Information Age*. London: Pluto Press. 2004.

Thompson, Emily. *The Soundscape of Modernity Architectural Acoustics and the Culture of Listening in America, 1900-1933*. Cambridge: MIT Press. 2002.

Vanderbeek, Stan. "Culture: Intercom and Expanded Cinema: A Proposal and Manifesto". *Film Culture*, No. 40, Spring. 1966. pp. 15-18.

Vanderbeek, Stan. "Movies: Disposable Art, Synthetic Media and Artificial Intelligence". *Take One Film Magazine*, Vol. 2, No. 3. 1969. pp. 14-16.

Voegelin, Salome. *Listening to Noise and Silence: Towards a Philosophy of Sound Art*. London: Continuum. 2010.

Walton, Kendall. *Mimesis as Make-Believe: On the Foundations of the Representational Arts*. London: Harvard University Press. 1990.

Warhol, Andy and Hackett, Pat. *POPism: The Warhol' 60s*. London: Hutchinson. 1980.

Weiss, Allen. S. *Mirrors of Infinity: The French Formal Garden and 17th Century Metaphysics*. New York: Princeton Architectural Press. 1995.

Weiss, Allen. S. *Phantasmic Radio*. Duke University Press. 1995.

Westernhagen, von, Curt. *Wagner: A Biography Volumes 1 and 2*. London: Cambridge University Press. 1978.

Wolff, Janet. *The Social Production of Art*. New York: New York University Press. 1993.

York, Herbert. "Nuclear Deterrence and the Military Uses of Space". *Daedalus: Weapons in Space Vol. I: Concepts and Technologies*. Issue 114, No. 2, Spring. 1985. pp. 17-32.

Young, La Monte. and Mac Low, Jackson. eds. *An Anthology of Chance Operations*. Bronx: Young and Mac Low. 1963.

Youngblood, Gene. *Expanded Cinema*. New York: E. P. Dutton and Co, Inc. 1970.

Additional Licenses

Figure 6: © 2006 Tessier, used under a Creative Commons Attribution-ShareAlike license: http://creativecommons.org/licenses/by-sa/3.0/

Figure 7: © 2007 Djtox The copyright holder of this file allows anyone to use it for any purpose, provided that the copyright holder is properly attributed. Redistribution, derivative work, commercial use, and all other use is permitted.

Figure 8: © 2006 Maurice Marcellin, used under a Creative Commons Attribution-ShareAlike license: http://creativecommons.org/licenses/by-sa/3.0/

Figure 9: © 2009 BobTheMad, used under a Creative Commons Attribution-ShareAlike license: http://creativecommons.org/licenses/by-sa/3.0/

Figure 10: © 2005 Andreas Praefcke, used under a Creative Commons Attribution-ShareAlike license: http://creativecommons.org/licenses/by-sa/3.0/

Figure 11: © 2005 Softeis, used under a Creative Commons Attribution-ShareAlike license: http://creativecommons.org/licenses/by-sa/3.0/

Figure 12: © 2004 Rafaelji, used under a Creative Commons Attribution-ShareAlike license: http://creativecommons.org/licenses/by-sa/3.0/

Figure 13: © 2007 tato grasso http://creativecommons.org/licenses/by-sa/2.5/

Figure 14: © 2005 Pabix, used under a Creative Commons Attribution-ShareAlike license: http://creativecommons.org/licenses/by-sa/3.0/

www.ingramcontent.com/pod-product-compliance
Lightning Source LLC
Chambersburg PA
CBHW031612210526
45464CB00004B/1542